Synthesizing Evidence:
The Art of Systematic Review

A Comprehensive Guide to Evidence Synthesis, Methodology, and Impact

MIHIR BHATTA

EDUTECH LEARNINGS, India

ISBN: 9798 3267 79670

Copyright © 2024 Mihir Bhatta
Published by EduTech Learnings Publishing House, India.

New Delhi, Kolkata, Pune, Asansol, Ranchi, Singapore.

All Rights Reserved. No part of this publication may be reproduced, stored in a retrieval system or transmitted in any form or by any means, electronic, mechanical, photocopying, recording, scanning or otherwise. Designations used by companies to distinguish their products are often claimed as trademarks. All brand names and product names used in this book are trade names, service marks, trademarks or registered trademarks of their respective owners. The Publisher is not associated with any product or vendor mentioned in this book. This publication is designed to provide accurate and authoritative information in regard to the subject matter covered.
It is sold on the understanding that the Publisher is not engaged in rendering professional services. If professional advice or other expert assistance is required, the services of a competent professional should be sought.

Acknowledgements

Writing "Synthesizing Evidence: The Art of Systematic Review" has been a deeply rewarding journey, and I like to thank all who made this journey possible.

First of all, I am thankful to the readers and scholars who have shown interest to my first book *"An Overview of Systematic Review and Meta-analysis"*. Your pursuit of knowledge and commitment to evidence-based practice inspire me to continue contributing to this important area of study. I am thankful to the researchers and students of University of Pretoria, South Africa, who graciously shared their knowledge and experiences with me. Your generosity and willingness to engage in dialogue were essential in enriching the content of this book.

I extend my heartfelt thanks to the faculties and staff of ICMR-NICED, Kolkata, whose insightful feedback and rigorous discussions significantly shaped the direction of this book. In particular, I would like to acknowledge Dr. Shanta Dutta (Scientist G and Director), Dr. Agniva Majumdar, Dr. Sandip Mukherjee, Dr. Debjit Chakraborty, Miss Piyali Ghosh, Dr. Ishanee Ghosal and Mr. Khokon Sen. I like to thank Dr. Subrata Biswas, Dr. Pradeep Kumar, Dr. Lalit Shakhawat of NACO. Dr. Rahul Biswas, WBSAPCS.

I also like to thank Dr. Pratap Thyran, Dr. Howard White, Dr. Anju Sinha Pradhan, Dr. Bhumika TV, Dr. BS Begapalli, Dr. SA Rizwan, Dr. Ana Pulimond, Dr. Komal Shah, Dr. TLN Prassad and Dr. Prabuddha Gopal Goshwami, from whom I have learned a lot. A special thanks to Dr. Rajatasubhra Adhikary for keeping faith on me also for his enormous encouragements.

Last but not the least, I am profoundly grateful to my father Mr. Manik C Bhatta, mother Purnima and sister Mou. Your unwavering support, patience, and encouragement provided the foundation upon which this work was built. To my wife, Manisha, your understanding and belief in my vision were invaluable during the late nights and long weekends spent writing and researching.

With gratitude,

Mihir

Content

	Page No.
Preface: The History of Systematic Review Tracing the Evolution from Past to Present	1
CHAPTER-1: Evidence Synthesis-Evidence map, organizations involved in evidence synthesis	5
CHAPTER-2: The importance of systematic review: advancing evidence-based practice and knowledge synthesis	23
CHAPTER-3: Types of reviews: understanding the differences between literature, narrative, scoping, and systematic reviews	31
CHAPTER-4: Evidence synthesis: methods and approaches for integrating research findings	43
CHAPTER-5: Developing a protocol for systematic review: A comprehensive guide	55
CHAPTER-6: Protocol registration for Systematic Review & Meta-Analysis	65
CHAPTER-7: Problem formulation, inclusion and exclusion criteria, and setting the scope in research	89
CHAPTER-8: Formation of review questions for Systematic Review	99

CHAPTER-9: Systematic methods of literature search — 105

CHAPTER-10: Search terms, sources of search, and e-databases — 111

CHAPTER-11: Documentation and reporting of searches — 125

CHAPTER-12: Screening potentially eligible studies: a comprehensive reporting standard — 131

CHAPTER-13: Grey literature: searching, inclusion, and significance — 143

CHAPTER-14: Data extraction and coding: techniques for effective data management and de-duplication — 149

CHAPTER-15: Critical appraisal and publication bias: ensuring rigor and addressing biases — 155

CHAPTER-16: Randomized controlled trials (RCTs): an advance in evidence-based practice — 193

CHAPTER-17: Disseminating Systematic Review findings: performing Evidence Synthesis — 205

CHAPTER-18: Technological advancement and applications of machine learning — 209

Questions and Answers	233
List of some legendary books	245
Bibliography	253
Annex 1: Sources of published literature:	295
Annex 2: Sources of Subject/Study Dependent published literature	296
Annex 3: Data extraction tool	297
Annex 4: PRISMA guidelines	304
Annex 5: Critical appraisal tools	321
Annex 6: Methods of search in different databases	331

Magnet

SYNTHESIZING EVIDENCE: THE ART OF SYSTEMATIC REVIEW: A COMPREHENSIVE GUIDE TO EVIDENCE SYNTHESIS, METHODOLOGY, AND IMPACT is a critical addition to the library of anyone involved in health research, clinical practice, or policy-making. its thorough exploration of systematic review methodology, combined with practical guidance and real-world examples, makes it an indispensable resource for ensuring that evidence synthesis is conducted with the highest standards of rigor and impact.

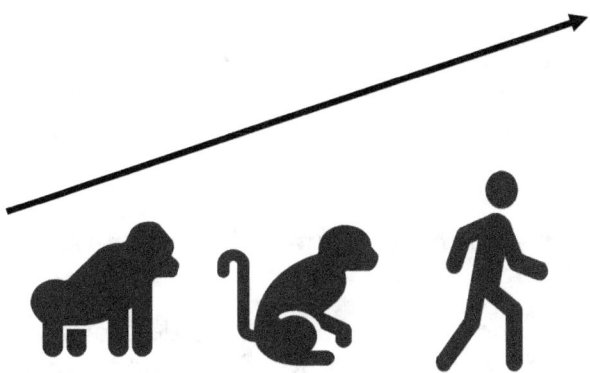

Preface

The History of Systematic Review - Tracing the Evolution from Past to Present

The voyage of systematic reviews is so fascinating, that reflects the evolution of evidence synthesis from its early stages to its current pivotal role in informing decision-making across various disciplines. Let's explore the historical timeline of systematic reviews in detail, highlighting key milestones and developments that have shaped their evolution over time.

Early Beginnings:

The concept of systematically synthesizing evidence dates back to ancient times, with scholars and philosophers attempting to organize knowledge and information in a structured manner. However, it wasn't until the latter part of the 20th century that the systematic review as we know it today began to take shape.

Archie Cochrane and Evidence-Based Medicine:

Archie Cochrane, a British epidemiologist, is often credited as one of the pioneers of evidence-based medicine and systematic reviews. In his influential book, "Effectiveness and Efficiency: Random Reflections on Health Services," published in 1972, Cochrane highlighted the importance of rigorous evaluation methods, including randomized controlled trials (RCTs), in assessing healthcare interventions. He emphasized the need for systematic reviews to synthesize evidence from multiple studies and guide clinical practice.

The Cochrane Collaboration:

The establishment of the Cochrane Collaboration in 1993 marked a significant milestone in the history of systematic reviews. Led by Iain Chalmers and others, the Cochrane Collaboration aimed to promote evidence-based healthcare by producing systematic reviews of healthcare interventions. The Collaboration developed standardized methods for conducting reviews, including comprehensive literature searches, critical appraisal of evidence, and transparent reporting of findings.

Methodological Advances:

Over the years, methodological advances have refined the process of conducting systematic reviews. The development of guidelines, such as the Preferred Reporting Items for Systematic Reviews and Meta-Analyses (PRISMA) statement, has improved the transparency and quality of systematic reviews. Methodological innovations, including network meta-analysis and qualitative synthesis techniques, have expanded the scope and applicability of systematic reviews across different research domains.

Expansion Across Disciplines:

While systematic reviews initially gained prominence in the field of healthcare and medicine, their utility has since expanded across various disciplines, including social sciences, education, environmental science, and public policy. Researchers in these fields recognize the value of systematic reviews in synthesizing evidence, informing policy decisions, and advancing knowledge.

Technological Innovations:

Advancements in technology have revolutionized the conduct and dissemination of systematic reviews. The development of systematic review software platforms, such as Covidence and

DistillerSR, has streamlined the review process, facilitating collaboration among researchers and improving efficiency. Online databases and repositories, such as PubMed, Cochrane Library, and PROSPERO, provide access to a vast array of systematic reviews, enhancing accessibility and dissemination of evidence.

Current Landscape and Future Directions:

Today, systematic reviews are widely recognized as a cornerstone of evidence-based practice and policymaking. They play a crucial role in synthesizing evidence, identifying gaps in knowledge, and guiding decision-making at individual, organizational, and policy levels. Looking ahead, the future of systematic reviews holds promise for further innovation, with ongoing efforts to address methodological challenges, enhance transparency and reproducibility, and integrate emerging technologies such as artificial intelligence and machine learning into the review process.

CHAPTER-1

EVIDENCE SYNTHESIS

Evidence synthesis is a systematic and rigorous process of collecting, evaluating, and integrating existing research to generate comprehensive and reliable conclusions on a particular topic or question. It is a critical aspect of evidence-based decision-making in various fields, including healthcare, social sciences, education, and public policy. The process of evidence synthesis typically involves the following steps:

Identifying the Research Question: The first step is to clearly define the research question or topic of interest. This helps to focus the synthesis on specific aspects and ensure relevant studies are included.

Literature Search: Researchers conduct a comprehensive search of existing literature, including published and unpublished studies, to identify all relevant research related to the question.

Study Selection: After the literature search, studies are screened based on predetermined inclusion and exclusion criteria. Only studies that meet specific quality and relevance standards are included in the synthesis.

Data Extraction: Researchers extract relevant data from the selected studies, including study characteristics, methodology, and findings.

Quality Assessment: The quality and risk of bias of each study are assessed to determine the strength and reliability of their findings.

Data Analysis: Depending on the nature of the studies, data may be statistically synthesized using meta-analysis or qualitatively summarized to identify common themes and patterns.

Interpretation and Conclusion: The evidence is synthesized and interpreted to generate meaningful conclusions and recommendations. The strength of the evidence and any limitations are also considered.

Dissemination: The results of the evidence synthesis are disseminated through reports, academic publications, or policy briefs to inform decision-makers, practitioners, and the public.

Evidence synthesis provides several benefits, including:

Reducing bias: By systematically identifying and evaluating all relevant studies, evidence synthesis minimizes the risk of bias and enhances the reliability of the conclusions.

Informing policy and practice: Synthesized evidence helps policymakers and practitioners make informed decisions based on the best available evidence.

Identifying research gaps: Through the process of evidence synthesis, gaps in existing research can be identified, guiding future research efforts.

Saving time and resources: Instead of duplicating individual studies, evidence synthesis builds upon existing research, making efficient use of resources.

Evidence synthesis plays a crucial role in advancing knowledge and facilitating evidence-based decision-making across various domains, ultimately leading to better-informed policies, practices, and interventions.

Process of an Evidence Synthesis:

Step 1: Formulating the Research Question
        Defining the Research Question
        Identifying PICO Elements (Population, Intervention, Comparison, Outcome)

Step 2: Literature Search
        Developing a Systematic Search Strategy
        Searching Electronic Databases and Grey Literature

Step 3: Study Selection
        Screening Search Results
        Applying Inclusion and Exclusion Criteria

Step 4: Data Extraction and Critical Appraisal
        Extracting Data from Selected Studies
        Assessing Study Quality and Risk of Bias

Step 5: Synthesis and Analysis
        Synthesizing Findings from Individual Studies
        Conducting Meta-Analysis (if applicable)

Step 6: Interpretation and Reporting
        Interpreting Synthesized Evidence
        Discussing Implications for Practice and Importance of Rigorous Methodology and Reporting

Evidence map

An evidence map, also known as a systematic map or scoping review, is a visual representation of the available evidence on a specific topic or research question. It is a systematic and structured approach to map out the existing literature, providing an overview of the volume, nature, and distribution of the evidence on the topic of interest. Unlike traditional systematic reviews or meta-analyses, which aim to synthesize and draw conclusions from the evidence, evidence maps focus on organizing and cataloguing the literature.

The process of creating an evidence map involves the following steps:

Define the research question: The first step is to clearly define the research question or topic of interest that the evidence map will address.

Conduct a comprehensive literature search: Researchers conduct a systematic search of the existing literature to identify all relevant studies related to the research question.

Screen and select studies: Similar to systematic reviews, studies are screened based on predetermined inclusion and exclusion criteria. The goal is to include studies that address the research question and meet specific quality standards.

Extract relevant data: Researchers extract key information from the selected studies, such as study characteristics, methodology, and main findings.

Organize the evidence: The extracted data is organized and summarized in a visual format, typically using tables, charts, or diagrams. The evidence map may categorize studies based on various criteria, such as

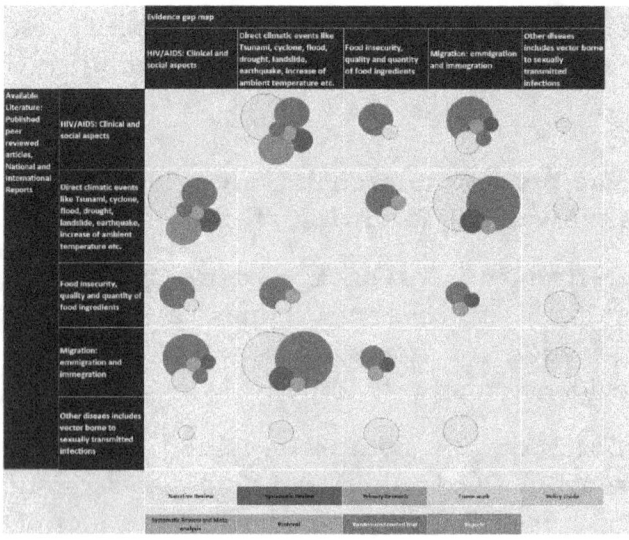

study design, geographic location, or intervention type.

Assess the quality and risk of bias: Some evidence maps include an assessment of the quality and risk of bias of the included studies to provide additional context to the findings. Interpret and report findings: The results of the evidence map are presented in a clear and transparent manner. The map may highlight research gaps, areas of consensus or controversy, and the overall state of the evidence on the topic.

Evidence maps are valuable tools for researchers, policymakers, and practitioners as they provide a snapshot of the available evidence, help identify research gaps, and inform the need for further investigation or more focused systematic reviews. They can be particularly useful when the research question is broad or when there is a large volume of literature on a particular topic that may not be feasible to synthesize comprehensively through a traditional systematic review.

Organizations involved in evidence synthesis

There are several organizations involved in evidence synthesis, which is the process of systematically collecting, evaluating, and synthesizing existing research to provide evidence-based recommendations. Some well-known organizations involved in evidence synthesis include:

Cochrane OR Cochrane Collaboration:

Cochrane is a global independent network of researchers, healthcare professionals, patients, and policymakers. It produces systematic reviews and meta-analyses of healthcare interventions and outcomes.

he Cochrane Collaboration, established in 1993, is a globally renowned independent organization dedicated to producing high-quality, credible evidence to inform healthcare decision-making. With thousands of contributors across more than 130 countries, including researchers, healthcare professionals, patients, and policymakers, Cochrane operates as a non-profit network focused on systematic reviews and evidence synthesis. Its core objectives encompass the production of rigorous systematic reviews, maintaining independence from commercial interests, and ensuring the integrity and credibility of its outputs. Through its decentralized structure of collaborative review groups, centers, and fields worldwide, Cochrane strives to provide evidence relevant to diverse populations and healthcare settings. Key contributions include its gold standard Cochrane Reviews, methodological innovation in systematic review methodologies, advocacy for evidence-based practice, and training and capacity building initiatives. Cochrane's impact extends to informing clinical guidelines, shaping healthcare policies, and empowering patients to make informed decisions about their health, thus solidifying its position as a trusted source of high-quality evidence for improving healthcare outcomes globally.

Key Contributions:

Cochrane Reviews: Cochrane reviews cover a wide range of healthcare topics, from treatments and interventions to diagnostic tests and public health measures. These reviews are considered gold standards in evidence-based healthcare and are widely used by clinicians, policymakers, and researchers.

Methodological Innovation: The Cochrane Collaboration has been at the forefront of developing and refining systematic review methodologies. It continually strives to improve methods for evidence synthesis, including statistical techniques, risk of bias assessment, and grading the quality of evidence.

Advocacy for Evidence-Based Practice: Through its publications, guidelines, and advocacy efforts, Cochrane promotes the importance of evidence-based practice in healthcare. It advocates for the use of high-quality evidence to inform clinical decision-making and healthcare policies worldwide.

Training and Capacity Building: Cochrane provides training and support for researchers and healthcare professionals interested in conducting systematic reviews and utilizing evidence-based practices. Its initiatives aim to build capacity for evidence synthesis and improve the quality of healthcare research globally.

Campbell Collaboration:

The Campbell Collaboration is an international organization that focuses on social science research synthesis. It conducts systematic reviews of the effects of social interventions in areas such as education, crime, social welfare, and more.

The Campbell Collaboration, founded in 2000, mirrors the Cochrane Collaboration's mission, albeit with a focus on social and behavioral sciences, education, and crime and justice. Like its counterpart, the Campbell Collaboration promotes evidence-based decision-making by producing systematic reviews and evidence synthesis in these fields. Operating as a non-profit network with contributors worldwide, including researchers, policymakers, practitioners, and stakeholders, the Campbell Collaboration emphasizes transparency, rigor, and independence in its work. Through its review groups and centers, it conducts rigorous evaluations of social interventions, educational programs, and crime prevention strategies, among others, to inform policy and practice. By providing accessible and credible evidence, the Campbell Collaboration aims to enhance the effectiveness and efficiency of social policies and interventions, contributing to positive social outcomes globally.

In 2000, a group of 85 social and behavioural scientists and social practitioners from 13 countries met in Philadelphia, USA and founded the Campbell Collaboration. The collaboration aims to address the need for an organisation that produces systematic reviews of research evidence on the effectiveness of social interventions. Many of the people involved in the establishment of the Campbell Collaboration were from Cochrane.

Joanna Briggs Institute (JBI):

JBI is an international research organization that develops evidence-based healthcare resources, including systematic reviews, evidence summaries, and best practice guidelines.

The Joanna Briggs Institute (JBI) is a renowned international research organization dedicated to promoting evidence-based healthcare practices through research, education, and dissemination. Established in Australia in 1996, JBI collaborates with healthcare professionals, researchers, and organizations worldwide to conduct rigorous systematic reviews, evidence syntheses, and guideline development in various healthcare disciplines.

JBI's primary mission is to bridge the gap between research evidence and clinical practice by providing high-quality evidence-based resources and tools. The institute offers a range of programs and services designed to support healthcare professionals and organizations in implementing evidence-based practices, including:

Systematic Reviews and Evidence Synthesis: JBI conducts systematic reviews and meta-analyses of research evidence to inform clinical practice, policy-making, and healthcare decision-making across a wide range of healthcare topics and disciplines.

Evidence-Based Practice Training: JBI provides comprehensive training programs, workshops, and online courses on evidence-based practice, systematic review methodology, and critical appraisal skills for healthcare professionals, researchers, and students.

Clinical Practice Guidelines: JBI develops evidence-based clinical practice guidelines to assist healthcare professionals in making informed decisions about patient care and treatment interventions, ensuring consistency and quality of care delivery.

Evidence-Based Healthcare Resources: JBI produces a range of evidence-based resources, including evidence summaries, best practice guidelines, and clinical decision support tools, to facilitate the implementation of evidence-based practices in healthcare settings.

Collaboration and Knowledge Translation: JBI collaborates with healthcare organizations, research institutions, and professional associations worldwide to promote knowledge translation and dissemination of evidence-based research findings.

By promoting evidence-based practice and facilitating the translation of research evidence into clinical practice, JBI plays a vital role in improving the quality, safety, and effectiveness of healthcare delivery globally. Its commitment to rigorous research methodologies, continuous education, and collaborative partnerships makes it a trusted resource for healthcare professionals seeking to enhance patient care and outcomes.

The International Organization for Standardization (ISO):

The International Organization for Standardization (ISO) conducts systematic reviews for its documents to ensure they remain relevant and up-to-date. Here's a breakdown of how the process works:

Review Frequency: ISO documents, like International Standards (IS) and Technical Specifications (TS), undergo systematic reviews at set intervals. IS documents are reviewed every 5 years, while TS documents are reviewed every 3 years. There's no fixed timeframe for Technical Reports (TR) reviews, but they can be requested by the Technical Committee (TC) at any time.

Feedback Gathering: During the review period, which typically lasts 20 weeks, all

ISO members are invited to provide feedback. TC members, especially those in TC 211, are required to participate. National standards bodies may also seek input from the public.

Voting Options: After the review, documents can be voted to confirm, amend, or withdraw. TS documents may also be voted to convert to IS.

Decision Making: Decisions are usually based on a simple majority vote. Further analysis may be conducted by the Programme maintenance group (PMG) and confirmed by a TC vote if necessary.

Next Steps: If a standard needs revision, project leaders and experts are appointed. If confirmed, the standard goes through another cycle before the next review. If withdrawn, there's an 8-week Withdraw Ballot.

Revision Process: Standards can be revised anytime by resolution in the TC, often proposed by the PMG or working groups.

Handling Concerns: If concerns arise during the review, they're discussed by the PMG and relevant working groups. If a revision is needed, it's proposed to the committee for decision.

PMG Guidance: PMG provides guidance based on reported issues with the standard, but it's not binding.

Review Outcomes: Outcomes vary, from unanimous confirmation to discussions on revision, withdrawal, or other actions based on majority vote.

For details, how Systematic Review conducted for International Standards: please consult *"Guidance on the Systematic Review process in ISO"*

Other major organization involved in evidence synthesis

Agency for Healthcare Research and Quality (AHRQ): AHRQ, a part of the U.S. Department of Health and Human Services, conducts and supports evidence synthesis and research on healthcare topics to improve patient outcomes and quality of care.

World Health Organization (WHO): WHO is a specialized agency of the United Nations that conducts evidence synthesis to inform global health policies and guidelines.

National Institute for Health and Care Excellence (NICE): NICE is a UK-based organization responsible for producing evidence-based guidance and recommendations for healthcare, public health, and social care.

Institute of Medicine (IOM): Now known as the National Academy of Medicine, the IOM conducts evidence synthesis and provides advice on health policy and practice in the United States.

Evidence Synthesis International (ESI): ESI is a consulting and training organization focused on providing evidence synthesis and systematic review services to clients in various fields.

These organizations play a crucial role in promoting evidence-based decision-making in healthcare, social sciences, and other sectors, helping to ensure that policies and practices are grounded in rigorous scientific evidence.

Collaboration for Environmental Evidence (CEE):

An open community of stakeholders working towards a sustainable global environment and the conservation of biodiversity. CEE seeks to promote and deliver evidence syntheses on issues of greatest concern to environmental policy and practice as a public service.

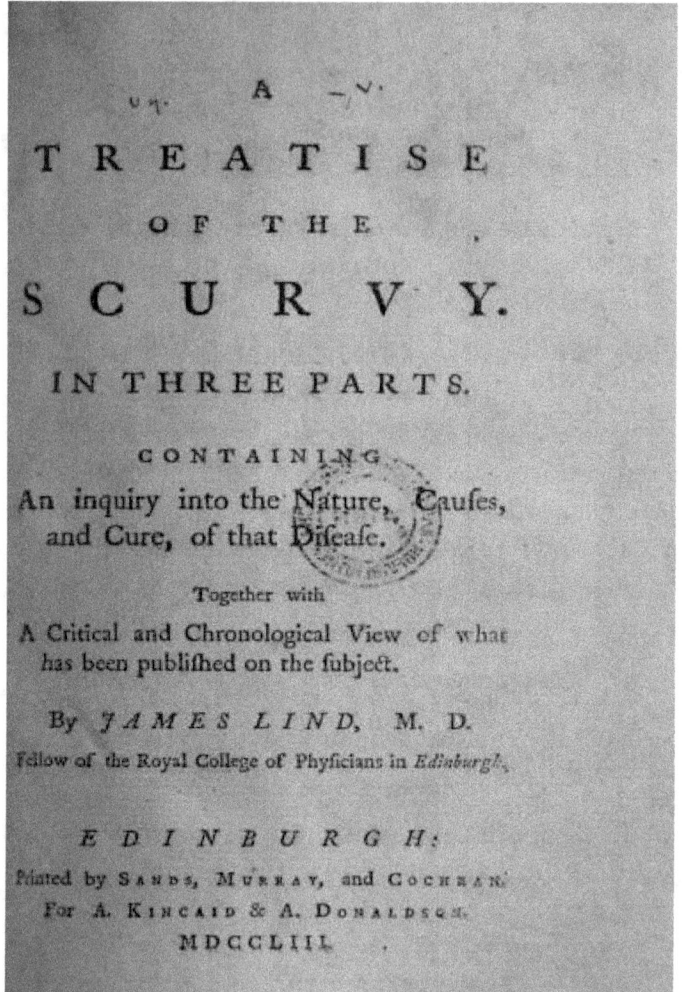

The first publication that is now recognized as equivalent to a modern systematic review was a 1753 publication by James Lind, which reviewed all of the previous publications about scurvy

CHAPTER-2

THE IMPORTANCE OF SYSTEMATIC REVIEW: ADVANCING EVIDENCE-BASED PRACTICE AND KNOWLEDGE SYNTHESIS

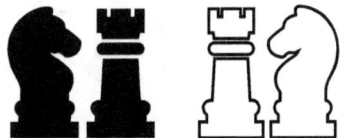

In the field of scientific research and evidence-based practice, the importance of systematic reviews cannot be overstated. Systematic reviews serve as essential tools for collating, evaluating, and synthesizing existing knowledge from various sources. These comprehensive and rigorous reviews contribute significantly to evidence-based decision-making, advancing knowledge, and guiding future research endeavours.

This note delves into the importance of systematic reviews, discussing their significance in promoting evidence-based practice, avoiding bias, enhancing research efficiency, and bridging gaps in knowledge. We will also explore the challenges and limitations associated with conducting systematic reviews.

What is a Systematic Review?

A systematic review is a methodical and transparent approach to gather, analyze, and interpret the results of multiple primary research studies on a specific topic. It involves a structured search of relevant literature, followed by a comprehensive evaluation and synthesis of the data to

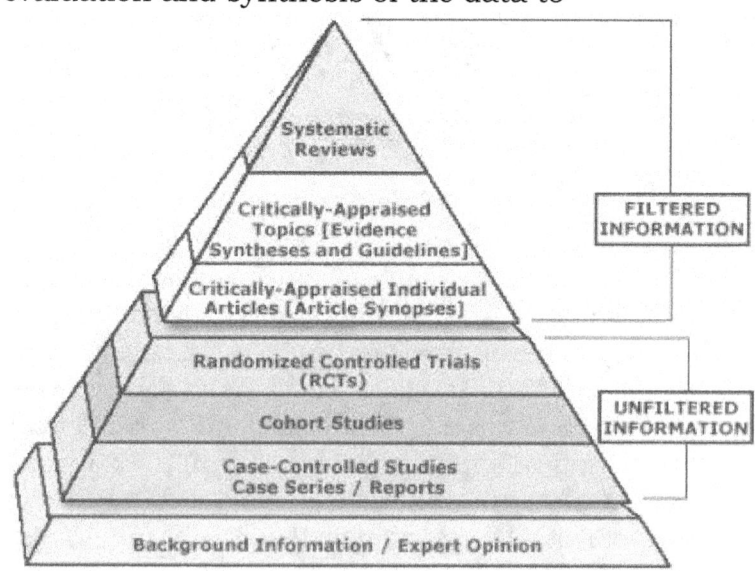

provide an objective and robust summary of the evidence available. Systematic reviews adhere to predefined criteria and methodologies, minimizing bias and subjectivity in the selection and interpretation of studies.

Advancing Evidence-Based Practice

Evidence-based practice (EBP) is a cornerstone of contemporary healthcare, education, and policy development. By synthesizing all available evidence, systematic reviews enable practitioners to make well-informed decisions. For instance, medical professionals can rely on systematic reviews of clinical trials to identify the most effective treatment options for specific conditions. Policymakers can utilize systematic reviews to shape evidence-based policies, informed by a comprehensive understanding of the available research on a particular topic. In essence, systematic reviews bridge the gap between research evidence and real-world application, enhancing the quality and efficiency of decision-making processes.

Avoiding Bias and Enhancing Transparency

One of the primary advantages of systematic reviews is their ability to minimize bias and subjectivity. The rigorous methodology involved ensures that all relevant studies, regardless of their outcomes, are included in the review. This approach reduces the risk of publication bias, where studies with positive or statistically significant results are more likely to be published, while negative or inconclusive findings may remain unpublished. By incorporating all available evidence, systematic reviews provide a more balanced and accurate representation of the research landscape.

Furthermore, systematic reviews enhance transparency in research. The detailed documentation of the search strategy, inclusion/exclusion criteria, and data synthesis methods allows other researchers to replicate the review and verify its findings. This transparency

Systematic reviews play a vital role in advancing evidence-based practice, guiding decision-making, and enhancing research efficiency. By consolidating and synthesizing existing knowledge, systematic reviews provide a holistic view of a research topic, identify knowledge gaps, and facilitate the formulation of evidence-based policies and practices. Despite the challenges and limitations, they face, systematic reviews remain indispensable tools for the pursuit of scientific knowledge, contributing significantly to progress and innovation across diverse fields. As technology and methodologies evolve, the importance of systematic reviews is likely to grow, shaping the future of evidence-based research and practice.

fosters a culture of accountability and credibility in scientific research.

Identifying Knowledge Gaps and Research Priorities

While individual studies contribute valuable insights, they often have limitations due to sample size, methodology, or context-specific factors. Systematic reviews help identify gaps in the existing knowledge base by highlighting inconsistencies, discrepancies, or areas with limited research. By pinpointing these gaps, systematic reviews can guide researchers towards new avenues of investigation, ensuring that future research efforts are focused on addressing the most pressing questions and producing meaningful contributions to the field.

Efficient Utilization of Resources

Conducting research is a resource-intensive endeavour, requiring significant investments in time, funding, and expertise. Systematic reviews optimize resource utilization by consolidating information from multiple studies. Instead of duplicating efforts on research questions that have already been explored, researchers can build upon existing systematic reviews to expand the scope of knowledge. This efficiency not only saves time and resources but also accelerates the pace of scientific advancement.

Holistic View and Meta-Analysis

Systematic reviews allow researchers to take a holistic view of a research topic. By synthesizing a wide range of studies, systematic reviews can reveal patterns, trends, and commonalities that may not be apparent in individual studies. Additionally, systematic reviews enable the use of meta-analysis, a statistical technique that combines data from multiple studies to produce more precise and robust estimates of effect sizes. Meta-analysis increases statistical power, providing a clearer understanding of the overall impact of an intervention or phenomenon.

Challenges and Limitations

Despite their numerous advantages, systematic reviews also face several challenges and limitations:

Time-Consuming Process*:* Conducting a systematic review can be a time-consuming endeavour, particularly for complex topics with a vast body of literature. The process of searching, screening, and analysing multiple studies can take several months to complete.

Resource Intensive*:* While systematic reviews optimize resource utilization in the long run, they still require initial investments in terms of skilled researchers, access to databases, and software tools.

Quality of Source Studies*:* The quality of a systematic review heavily relies on the quality of the primary studies included. If the primary studies have limitations, such as biases or methodological flaws, the systematic review's conclusions may also be compromised.

Publication Bias: Although systematic reviews aim to minimize publication bias, it may still be present if relevant studies with negative or null results are not published or available in the literature.

Inherent Heterogeneity: Studies included in a systematic review may vary in terms of methodology, population, interventions, and outcomes, introducing heterogeneity that may complicate data synthesis.

Availability of Data: Sometimes, researchers encounter difficulties in accessing the complete data from all the studies, hindering a comprehensive analysis.

Different field where Systematic Review is possible are as follows:

Environment
Business
Criminology
Education
Health sciences and medicine
Law
Psychology
Social work

CHAPTER-3

TYPES OF REVIEWS: UNDERSTANDING THE DIFFERENCES BETWEEN LITERATURE, NARRATIVE, SCOPING, AND SYSTEMATIC REVIEWS

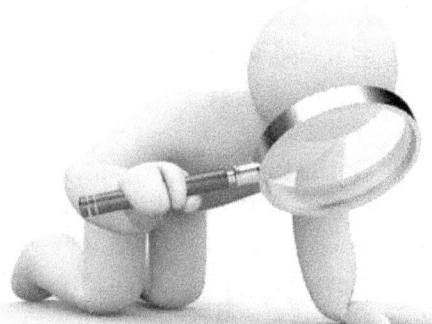

In the world of academic research and evidence-based practice, reviews are critical tools for understanding the existing knowledge on a specific topic or research question. Different types of reviews serve distinct purposes, ranging from providing an overview of the literature to conducting comprehensive and rigorous syntheses of evidence. This note delves into the various types of reviews, namely literature reviews, narrative reviews, scoping reviews, and systematic reviews, highlighting their characteristics, methodologies, and contributions to knowledge synthesis.

1. Literature Review

Definition and Purpose

A literature review is a critical examination and analysis of published scholarly works, research papers, books, and other relevant sources on a particular topic or research question. The primary purpose of a literature review is to provide a comprehensive overview and summary of the existing literature related to the chosen subject. It aims to identify key themes, trends, gaps, and controversies in the field, offering a broader understanding of the topic's current state of knowledge.

Characteristics

Inclusion Criteria: Literature reviews may include both primary and secondary sources, such as research articles, review papers, and theoretical contributions. The inclusion criteria may vary, depending on the scope and objectives of the review.

Search Strategy: While literature reviews require a systematic approach to searching for relevant sources, they are generally less structured than systematic reviews. Researchers use keywords, subject headings, and bibliographic databases to find relevant literature.

Data Synthesis: Literature reviews use qualitative or narrative synthesis methods to summarize and analyze the findings of the included studies. Researchers often present the data thematically or chronologically, highlighting patterns and key findings.

Scope: Literature reviews can be broad or narrow in scope, depending on the research question and available literature. They may cover a wide range of topics or focus on a specific aspect of the subject.

Advantages and Limitations

Literature reviews are useful for gaining an overall understanding of a research area and identifying gaps in knowledge. They are relatively less time-consuming compared to systematic reviews, making them suitable for preliminary investigations or exploring emerging fields. However, the lack of a standardized methodology may lead to potential bias in study selection and data synthesis.

2. Narrative Review

Definition and Purpose

A narrative review, also known as a traditional review or qualitative review, is a descriptive and interpretive summary of the literature on a particular topic. Unlike systematic reviews, narrative reviews do not follow a strict methodology, systematic search process, or predefined inclusion criteria. Instead, they rely on the author's expertise and interpretation of the literature to present an overview of the subject.

Characteristics

Search Strategy: Narrative reviews typically involve a less structured approach to literature search compared to systematic reviews. Authors often draw on their knowledge of the field and may reference key or influential papers.

Data Synthesis: The synthesis of data in a narrative review is qualitative and interpretive, rather than statistical. Authors use their judgment to identify key themes, concepts, and findings in the literature.

Scope: Narrative reviews may have a broad scope, encompassing a wide range of literature, or focus on a specific aspect of the subject, depending on the author's preferences.

Advantages and Limitations

Narrative reviews are relatively easier and quicker to produce compared to systematic reviews. They allow authors to present their expert opinions and interpretations, making them suitable for fields where data may be limited or preliminary. However, the lack of systematic methods may introduce bias and reduce the transparency of the review process.

3. Scoping Review

Definition and Purpose

A scoping review is a comprehensive and systematic approach to mapping and exploring the existing literature on a particular research question or topic. The primary purpose of a scoping review is to identify the breadth and depth of the available evidence, as well as gaps in the research landscape. Scoping reviews are especially useful when the research question is broad, complex, or multi-faceted.

Characteristics

Inclusion Criteria: Scoping reviews typically have broader inclusion criteria compared to systematic reviews. They aim to capture a wide range of literature to map the extent and diversity of existing research.

Search Strategy: Scoping reviews follow a systematic and transparent search process, similar to systematic reviews, although they may include a wider range of sources, such as gray literature and unpublished studies.

Data Synthesis: Data synthesis in a scoping review involves a descriptive and thematic analysis of the literature. Researchers identify key themes and concepts, categorize the evidence, and present an overview of the research landscape.

Scope: Scoping reviews are especially valuable for exploring emerging or complex research topics, where the boundaries of the literature are not well-defined.

Advantages and Limitations

Scoping reviews provide a comprehensive overview of the existing literature, helping researchers understand the breadth and depth of a research area. They are particularly useful for identifying research gaps and potential areas for future investigation. However, scoping reviews may not provide the same level of detail or critical appraisal as systematic reviews, limiting their ability to draw definitive conclusions about the effectiveness of interventions or phenomena.

4. Systematic Review

Definition and Purpose

A systematic review is a rigorous and comprehensive approach to synthesizing the evidence from multiple primary studies on a specific research question. It involves a structured and transparent process, adhering to predefined criteria and methods, to minimize bias and subjectivity. Systematic reviews are considered the gold standard for evidence synthesis, providing the highest level of confidence in the findings.

Characteristics

Inclusion Criteria: Systematic reviews have well-defined inclusion and exclusion criteria to ensure that only relevant and high-quality studies are included. These criteria are established before the review begins to minimize the risk of bias.

Methodology: Systematic reviews follow a strict and standardized methodology, including predefined search strategies, inclusion criteria, and data synthesis techniques. On the other hand, literature and narrative reviews are less structured, relying on the author's expertise and interpretation of the literature. Scoping reviews are systematic but have broader inclusion criteria and focus on mapping the research landscape.

Search Strategy: Systematic reviews follow a detailed and systematic search process to identify all relevant studies. Researchers use a combination of keywords, subject headings, and database filters to ensure the comprehensive retrieval of literature.

Data Synthesis: Systematic reviews use statistical methods, such as meta-analysis, to combine data from multiple studies. Literature and narrative reviews use qualitative synthesis to summarize findings thematically or descriptively. Scoping reviews also employ qualitative synthesis but aim to map the evidence rather than draw conclusions about effectiveness.

Scope: Systematic reviews are focused on answering specific research questions with high levels of evidence. Literature reviews can have broad or narrow scopes, while narrative reviews may lack clear boundaries. Scoping reviews aim to map the breadth and depth of research on a topic, making them suitable for exploring emerging or complex areas.

Transparency and Bias: Systematic reviews are highly transparent, with a clear and replicable methodology, minimizing bias in study selection and data synthesis. Literature and narrative reviews may have less transparent methods, potentially introducing bias. Scoping reviews are transparent in their search process but may have broader inclusion criteria, leading to some level of bias.

Time and Resources: Systematic reviews are the most resource-intensive and time-consuming type of review due to their rigorous methodology. Literature and narrative reviews are generally quicker and require fewer resources. Scoping reviews also require significant effort but may be less demanding than systematic reviews.

> **In summary, different** types of reviews serve distinct purposes in the research and evidence-based practice landscape. Literature reviews provide overviews of the existing literature, narrative reviews offer expert interpretations, scoping reviews map the research landscape, and systematic reviews provide the highest level of evidence through rigorous methodologies and statistical analyses. Each type of review has its advantages and limitations, and researchers should choose the most appropriate type based on their research question, available resources, and desired level of evidence. By employing the right type of review, researchers can contribute to knowledge synthesis and evidence-based decision-making in their respective fields.

Advantages and Limitations

Systematic reviews are regarded as the most reliable and informative type of review due to their rigorous methodology and comprehensive approach. They offer a high level of evidence for decision-making, policy formulation, and clinical practice. However, conducting a systematic review requires significant time, expertise, and resources, making them more demanding than other types of reviews.

5. Umbrella reviews

In medical research, an umbrella review is a review of systematic reviews or meta-analyses.] They may also be called overviews of reviews, reviews of reviews, summaries of systematic reviews, or syntheses of reviews. Umbrella reviews are among the highest levels of evidence currently available in medicine.

By summarizing information from multiple overview articles, umbrella reviews make it easier to review the evidence and allow for comparison of results between each of the individual reviews. Umbrella reviews may address a broader question than a typical review, such as discussing multiple different treatment comparisons instead of only one. They are especially useful for developing guidelines and clinical practice, and when comparing competing interventions

Archibald Leman Cochrane

Archibald Leman Cochrane (12 January 1909 – 18 June 1988) was a Scottish physician noted for his book, Effectiveness and Efficiency: Random Reflections on Health Services, which advocated the use of randomized controlled trials (RCTs) to improve clinical trials and medical interventions. His advocacy of RCTs eventually led to the creation of the Cochrane Library database of systematic reviews, the UK Cochrane Centre in Oxford and Cochrane (previously known as the Cochrane Collaboration), an international organization of review groups that are based at research institutions worldwide. He is known as one of the fathers of modern clinical epidemiology and is considered to be the originator of the idea of evidence-based medicine. The Archie Cochrane Archive is held at the Archie Cochrane Library at University Hospital Landough, Penarth.

Cochrane was appointed David Davies Professor of Tuberculosis and Chest Diseases at the Welsh National School of Medicine, now Cardiff University School of

Medicine in 1960. Nine years later he became Director of the new Medical Research Council's Epidemiology Research Unit in Cardiff. His ground-breaking paper on validation of medical screening procedures, published jointly with fellow epidemiologist Walter W. Holland in 1971, became a classic in the field. His 1971 Rock Carling Fellowship Monograph Effectiveness and Efficiency: Random Reflections on Health Services, first published in 1972 by the Nuffield Provincial Hospitals Trust, now known as the Nuffield Trust, was very influential.

Maintaining this challenge to the medical care system as he saw it, in 1978, with colleagues, he published a study of 18 developed countries in which he made the observations at box.

> *the indices of health care are not negatively associated with mortality, and there is a marked positive association between the prevalence of doctors and mortality in the younger age groups. No explanation of this doctor anomaly has so far been found. Gross national product per head is the principal variable which shows a consistently strong negative association with mortality*

This work was selected for inclusion in a compendium of influential papers, from historically important epidemiologists, published by the Pan American Health Organization (PAHO/WHO) in 1988. Cochrane promoted the randomised trial and is a co-author with Professor Peter Elwood on a report on the first randomised trial of aspirin in the prevention of vascular disease.

LEGEND

CHAPTER-4

EVIDENCE SYNTHESIS: METHODS AND APPROACHES FOR INTEGRATING RESEARCH FINDINGS

In the realm of evidence-based practice and policy-making, evidence synthesis plays a crucial role in integrating research findings to inform decision-making processes. Evidence synthesis involves the systematic and rigorous process of combining, summarizing, and analyzing data from multiple studies on a specific topic. By synthesizing evidence, researchers can draw more robust conclusions, identify patterns, and make informed recommendations. This note explores the different methods of evidence synthesis, including narrative synthesis, meta-analysis, systematic reviews, mixed-methods synthesis, and realist synthesis, highlighting their characteristics, advantages, and limitations.

1. **Narrative Synthesis**

Definition and Purpose

Narrative synthesis is a qualitative method of evidence synthesis that involves summarizing and interpreting research findings in a narrative format. Unlike quantitative methods, narrative synthesis does not involve statistical analysis or meta-analysis. Instead, it relies on a descriptive approach to presenting and discussing the results of individual studies, aiming to identify themes, patterns, and discrepancies within the research.

Characteristics

Data Presentation: Narrative synthesis presents the results of individual studies in the form of textual summaries, tables, or figures. Researchers may use quotes, paraphrases, or qualitative codes to illustrate key findings.

Analysis: Researchers interpret the findings of individual studies and compare and contrast their results to identify common themes or divergent findings.

Scope: Narrative synthesis can be used for both qualitative and quantitative studies, making it versatile in accommodating various types of evidence.

Advantages and Limitations

Narrative synthesis is particularly useful when dealing with diverse types of evidence or when meta-analysis is not feasible due to heterogeneity among the included studies. It allows for a more in-depth exploration of the research findings and helps highlight contextual factors that may influence the results. However, narrative synthesis is susceptible to potential bias and subjectivity, as it relies heavily on the researcher's interpretations and judgments.

2. Meta-Analysis

Definition and Purpose

Meta-analysis is a quantitative method of evidence synthesis that involves pooling data from multiple primary studies to produce a single combined estimate of the effect size. It is widely used in systematic reviews and is particularly effective in providing more precise and reliable estimates of the treatment effect.

Characteristics

Effect Size Calculation: Meta-analysis involves calculating effect sizes (e.g., mean differences, odds ratios, risk ratios) for each study and then combining these effect sizes using statistical methods.

Weighting: Studies with larger sample sizes or lower variability are given more weight in the meta-analysis, as they are considered more reliable and contribute more to the overall estimate.

Heterogeneity Assessment: Meta-analyses often include tests for heterogeneity to assess the variability of effect sizes across studies. If significant heterogeneity is present, researchers may explore potential sources of variation through subgroup analyses or meta-regression.

Advantages and Limitations

Meta-analysis provides a quantitative and objective summary of research findings, increasing statistical power and precision. It can help identify patterns and trends that may not be evident in individual studies. However, meta-analysis requires strict adherence to methodological standards, and the quality of the overall estimate is heavily dependent on the quality of the included studies. Moreover, it may not be suitable when the included studies are too dissimilar or have substantial heterogeneity.

3. Systematic Reviews

Definition and Purpose

Systematic reviews are comprehensive and structured reviews that aim to synthesize the evidence on a specific research question or topic. They integrate multiple research studies using rigorous and transparent methods to minimize bias and subjectivity.

Characteristics

Inclusion Criteria: Systematic reviews have predefined inclusion and exclusion criteria to select relevant studies based on the research question.

Search Strategy: Systematic reviews follow a detailed and systematic search process to identify all relevant studies. Researchers use a combination of keywords, subject headings, and database filters to ensure the comprehensive retrieval of literature.

Data Synthesis: Systematic reviews use both qualitative and quantitative synthesis methods, such as narrative synthesis and meta-analysis, to summarize and analyze the findings of the included studies.

Scope: Systematic reviews focus on specific research questions and aim to provide evidence-based answers to these questions.

Advantages and Limitations

Systematic reviews are considered the gold standard in evidence synthesis, providing the highest level of confidence in research findings. They are transparent and reproducible, facilitating the integration of evidence into decision-making processes. However, conducting a systematic review requires significant time, expertise, and resources, making them more demanding than other types of evidence synthesis.

4. **Mixed-Methods Synthesis**

Definition and Purpose

Mixed-methods synthesis is an approach that combines both qualitative and quantitative evidence to answer research questions more comprehensively. It involves integrating findings from qualitative studies (e.g., through narrative synthesis) and quantitative studies (e.g., through meta-analysis) to provide a more holistic understanding of the topic under investigation.

Characteristics

Data Integration: Mixed-methods synthesis combines qualitative and quantitative data in a single review, allowing for a deeper exploration of the research question.

Inclusion Criteria: Mixed-methods synthesis may include both qualitative and quantitative studies, or it may involve separate syntheses of each type of evidence followed by an integration of the results.

Scope: Mixed-methods synthesis is suitable for research questions that require a multifaceted approach or when evidence from different types of studies is available.

Advantages and Limitations

Mixed-methods synthesis is beneficial in capturing both the breadth and depth of evidence on a research question. It provides a more comprehensive understanding of the topic by integrating diverse types of evidence. However, conducting mixed-methods synthesis requires expertise in both qualitative and quantitative research methods, making it more complex and challenging to implement.

5. **Realist Synthesis**

Definition and Purpose

Realist synthesis is an approach to evidence synthesis that focuses on understanding "what works for whom, under what circumstances, and why." It aims to uncover underlying mechanisms and contextual factors that explain how interventions or programs produce outcomes.

Characteristics

Contextual Analysis: Realist synthesis emphasizes context and seeks to identify the mechanisms by which interventions interact with specific contexts to produce outcomes.

Theory Development: Realist synthesis often involves developing or refining theoretical models that explain the causal relationships between interventions and outcomes.

Scope: Realist synthesis is particularly useful when investigating complex interventions or when there is a need to understand the interactions between interventions and contexts.

Advantages and Limitations

Realist synthesis provides valuable insights into the underlying mechanisms and contextual factors that influence the effectiveness of interventions. It is well-suited for addressing complex research questions and can inform the development of more effective interventions. However, realist synthesis requires a deep understanding of the subject matter and the ability to develop and test theoretical models, making it a more challenging approach compared to other evidence synthesis methods.

Evidence synthesis encompasses a diverse range of methods and approaches, each tailored to specific research questions and objectives. Narrative synthesis offers a qualitative exploration of the literature, while meta-analysis provides quantitative estimates of effect sizes. Systematic reviews combine both qualitative and quantitative methods to provide evidence-based answers to research questions. Mixed-methods synthesis integrates diverse types of evidence, and realist synthesis emphasizes understanding the context and mechanisms that underlie intervention effectiveness.

Researchers and policymakers should carefully select the most appropriate method of evidence synthesis based on the available evidence, research question complexity, and desired level of evidence-based guidance. By employing rigorous and transparent methods of evidence synthesis, researchers can contribute to advancing knowledge, informing decision-making, and promoting evidence-based practice across various fields.

Types of systematic reviews

Systematic reviews can vary in their scope, methodology, and focus based on the research question, objectives, and inclusion criteria. Here are some different types of systematic reviews:

Intervention/Systematic Reviews of Interventions:

These reviews focus on evaluating the effectiveness of interventions, treatments, therapies, or preventive measures for specific health conditions or outcomes. They typically assess randomized controlled trials (RCTs) and other comparative studies to determine the impact of interventions on patient outcomes.

Diagnostic Accuracy/Systematic Reviews of Diagnostic Tests:

These reviews aim to assess the accuracy and reliability of diagnostic tests or screening measures for detecting specific health conditions or diseases. They often evaluate studies that compare the performance of diagnostic tests against a reference standard, such as gold standard tests or clinical diagnosis.

Prognostic/Systematic Reviews of Prognostic Factors:

These reviews examine the predictive value of prognostic factors or predictors for certain health outcomes or disease progression. They synthesize evidence from longitudinal studies to identify factors associated with better or worse prognoses for individuals with particular conditions.

Qualitative/Systematic Reviews of Qualitative Studies:

These reviews focus on synthesizing qualitative research findings to explore complex phenomena, experiences, or perspectives related to health and healthcare. They involve the systematic identification, appraisal, and synthesis of qualitative studies, such as interviews, focus groups, or ethnographic research.

Economic/Systematic Reviews of Economic Evaluations:

These reviews analyze economic evaluations, cost-effectiveness analyses, or cost-benefit studies related to healthcare interventions or policies. They assess the economic implications of healthcare interventions and help policymakers make informed decisions about resource allocation and healthcare financing.

Mixed-Methods/Systematic Reviews of Mixed Methods Studies:

These reviews integrate both quantitative and qualitative evidence to address research questions that require a comprehensive understanding of complex phenomena. They combine data from diverse study designs and methodologies to provide a more holistic perspective on the topic of interest.

Network/Systematic Reviews of Network Meta-Analysis:

These reviews compare multiple interventions using both direct and indirect evidence from randomized controlled trials. They allow for the simultaneous comparison of multiple treatments and estimation of their relative effectiveness, even when head-to-head comparisons are limited.

Scoping/Systematic Scoping Reviews:

These reviews aim to map the breadth and depth of research literature on a particular topic, identifying key concepts, evidence gaps, and research trends. They provide an overview of the existing literature to inform the development of research agendas or policy initiatives.

CHAPTER-5

DEVELOPING A PROTOCOL FOR SYSTEMATIC REVIEW: A COMPREHENSIVE GUIDE

Systematic reviews are rigorous and transparent approaches to synthesizing evidence from multiple studies on a specific research question. The development of a well-defined protocol is a crucial step in conducting a systematic review. A protocol outlines the objectives, methods, and procedures that researchers will follow throughout the review process. It helps ensure that the review is conducted systematically, minimizing bias, and increasing the reliability and validity of the findings. This note provides a comprehensive guide on how to develop a protocol for a systematic review, covering the key components and considerations involved in the process.

Define the Research Question

The first step in developing a protocol for a systematic review is to define the research question clearly. The research question should be specific, focused, and relevant to the field of study. It should also be formulated using the PICO (Population, Intervention, Comparison, Outcome) framework to guide the search for relevant studies. The PICO framework consists of four key elements:

Population: The specific group of individuals or participants to be studied.

Intervention: The intervention or exposure being investigated.

Comparison: The alternative or control group, if applicable.

Outcome: The outcomes of interest that will be measured or assessed.

A clear research question provides the foundation for the entire review process, guiding the search strategy, study selection, and data synthesis.

Guidelines and standards for systematic review

There are several guidelines and standards available for conducting systematic reviews and meta-analyses across various disciplines. These guidelines provide researchers with a framework for designing, conducting, and reporting systematic reviews in a rigorous and transparent manner. Here are some of the most commonly used guidelines:

PRISMA (Preferred Reporting Items for Systematic Reviews and Meta-Analyses): PRISMA is one of the most widely recognized guidelines for reporting systematic reviews and meta-analyses of healthcare interventions. It consists of a 27-item checklist and a flow diagram to ensure transparent reporting of key methodological details and findings.

Cochrane Handbook for Systematic Reviews of Interventions: Produced by the Cochrane Collaboration, this handbook provides detailed guidance on conducting systematic reviews and meta-analyses of healthcare interventions. It covers all aspects of the review process, from formulating research questions to synthesizing evidence and interpreting findings.

MOOSE (Meta-analysis of Observational Studies in Epidemiology): MOOSE is a set of guidelines for reporting meta-analyses of observational studies in epidemiology. It provides recommendations for transparent reporting of key methodological details, such as study selection, data extraction, and statistical analysis.

STROBE (STrengthening the Reporting of OBservational studies in Epidemiology): STROBE is a set of guidelines for reporting observational studies in epidemiology. While not specific to systematic reviews, STROBE provides recommendations for transparent reporting of observational research, which is often included in systematic reviews as part of the evidence synthesis process.

ENTREQ (Enhancing Transparency in Reporting the Synthesis of Qualitative Research): ENTREQ is a set of guidelines for reporting systematic reviews and meta-syntheses of qualitative research. It provides recommendations for transparent reporting of key methodological details and findings in qualitative evidence synthesis.

Conduct a Scoping Search

Before finalizing the research question, it is beneficial to conduct a scoping search to assess the availability of relevant studies. The scoping search helps determine if there is a sufficient number of studies on the chosen topic and identifies potential challenges in finding appropriate evidence. This preliminary search may also reveal any gaps in the literature or areas with limited research. The findings from the scoping search can inform the final formulation of the research question and help researchers set realistic expectations for the review.

Assemble the Review Team

A systematic review is a time-consuming and complex endeavor that requires a multidisciplinary team with diverse expertise. The review team typically includes researchers, content experts, methodologists, statisticians, and information specialists (librarians). Each team member should have a clear understanding of their roles and responsibilities during the review process. Collaborative teamwork ensures that the review is comprehensive, well-informed, and transparent.

Develop the Protocol Structure

The next step is to develop the structure of the protocol document. The protocol should be well-organized and include the following key components:

• Title and Abstract: A concise and informative title that reflects the research question and scope of the review. The abstract should provide a brief overview of the review's objectives, methods, and expected outcomes.

• Background and Rationale: A clear justification for the review, explaining the importance and relevance of the research question. The background should also include a summary of existing knowledge and any gaps in the literature.

- Objectives: Clearly state the primary and secondary objectives of the systematic review. Objectives should be specific and aligned with the research question.

- Methods: Describe the methods that will be used in the review, including the study design, data sources, search strategy, study selection criteria, data extraction methods, and data synthesis approach (qualitative, quantitative, or mixed-methods).

- Quality Assessment: Outline the criteria and methods for assessing the quality and risk of bias in the included studies. This may include tools such as the Cochrane Risk of Bias Tool for randomized controlled trials or the Newcastle-Ottawa Scale for observational studies.

- Data Synthesis: Describe the planned approach for synthesizing the data from the included studies. This may involve meta-analysis, narrative synthesis, or both, depending on the nature of the data and research question.

- Publication Bias: Detail the steps to address publication bias, such as searching for unpublished studies or grey literature.

- Ethical Considerations: Address any ethical considerations related to the review, such as patient confidentiality or the need for ethical approval.

- Timeline and Resources: Provide a timeline for the completion of the review and outline the resources required, including personnel, funding, and access to databases and software.

- Dissemination Plan: Describe how the findings of the review will be disseminated, such as through academic publications, conferences, or policy briefs.

Develop the Search Strategy

The search strategy is a critical component of the systematic review, as it determines the scope and comprehensiveness of the evidence retrieved. The search strategy should be designed to identify all relevant studies on the research question while minimizing the risk of missing key articles. It typically involves a combination of keywords, subject headings (MeSH terms), and database filters specific to each bibliographic database to be searched.

The information specialist or librarian on the review team plays a vital role in developing an effective search strategy. They can assist in identifying relevant databases, designing the search terms, and conducting the actual searches.

Define Study Selection Criteria

Study selection criteria are essential to ensure that the review includes only studies that meet specific eligibility criteria. These criteria should be pre-defined and based on the research question and the PICO framework. Researchers should consider the following elements when defining study selection criteria:

• Study Design: Specify the types of studies to be included (e.g., randomized controlled trials, observational studies, qualitative studies).

• Population: Define the characteristics of the study population, including age, gender, health condition, or other relevant factors.

• Intervention: Clearly define the intervention(s) or exposure(s) of interest.

• Comparison: If applicable, specify the comparison group to be considered in the review.

• Outcome: Define the primary and secondary outcomes that will be assessed in the review.

Pilot Test the Protocol

Before proceeding with the full review, it is advisable to pilot test the protocol on a small subset of studies. This pilot test helps identify any ambiguities or challenges in the protocol and provides an opportunity to refine the methodology and inclusion criteria.

Registering the Protocol

Registering the protocol in a public repository, such as PROSPERO (International Prospective Register of Systematic Reviews), is highly recommended. Protocol registration helps prevent duplication of effort, increases transparency, and provides a record of the review's

Once, the protocol is finalized and registered, researchers can proceed with the systematic review following the defined methods and procedures.

Developing a protocol is a critical and systematic process that lays the groundwork for conducting a rigorous and transparent review. A well-defined protocol ensures that the review follows a structured and unbiased approach, increasing the reliability and validity of the findings. By adhering to the steps outlined in this comprehensive guide, researchers can successfully develop a robust protocol that guides the systematic review process and contributes to evidence-based practice and decision-making in various fields.

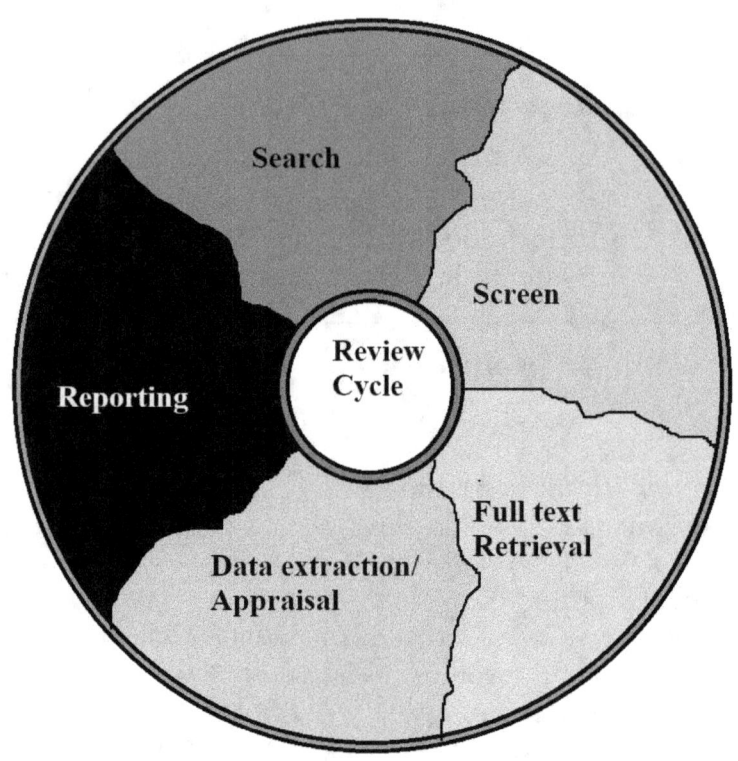

Review cycle

CHAPTER-6

PROTOCOL REGISTRATION FOR SYSTEMATIC REVIEW & META-ANALYSIS

A research protocol is a comprehensive document that outlines the design, methods, and objectives of a study. Registering a research protocol involves submitting this document to a publicly accessible registry before commencing the study.

This chapter explores the importance of protocol registration in various fields of research, its benefits, the process of registration, and the impact it has on research integrity and evidence-based practice.

6.1. The Significance of Protocol Registration

6.1.1. Enhancing Research Transparency

Protocol registration promotes transparency by making research plans publicly available, ensuring that researchers cannot alter the research question or methodology post hoc.

6.1.2. Preventing Publication Bias

Protocol registration prevents publication bias by reducing the likelihood of selective reporting, where only studies with favourable outcomes are published.

6.1.3. Reducing Duplicate Research

By making research plans accessible, protocol registration reduces the risk of duplication, saving resources and time.

6.1.4. Strengthening Research Integrity

Protocol registration fosters research integrity by promoting a clear distinction between planned and unplanned analyses and outcomes.

6.2. Common Protocol Registration Platforms

6.2.1. ClinicalTrials.gov

ClinicalTrials.gov is a widely used registry for clinical trials, maintaining a comprehensive database of both ongoing and completed trials across various therapeutic areas.

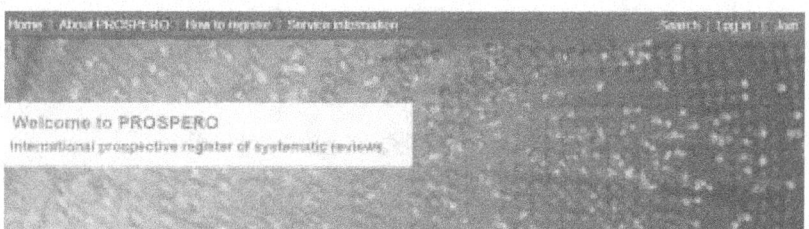

6.2.2. PROSPERO

PROSPERO is an international prospective register of systematic reviews, providing a platform for researchers to register systematic review protocols.

6.2.3. Other Discipline-Specific Registries

Various disciplines have their own protocol registration platforms, catering to specific research areas like public health, social sciences, and environmental scienc

6.3.1. Background and Rationale

The background provides context for the study and outlines the rationale for conducting the research.

6.3.2. Objectives

Clearly state the primary and secondary objectives of the study.

6.3.3. Methods

Detail the study design, sample size calculation, data collection methods, and data analysis plan.

6.3.4. Ethical Considerations

Address ethical considerations and describe how the study will protect the rights and welfare of participants.

6.3.5. Funding and Resources

Include information about funding sources and the resources available to conduct the research.

Protocol Reporting Guidelines

MECIR (Methodological Expectations for Cochrane Intervention Reviews) Manual - guidelines on reporting protocols for Cochrane Intervention reviews

PRISMA-P - (PRISMA (Preferred Reporting Items for Systematic Reviews) for systematic review protocols

6.4. Protocol Registration Process

6.4.1. Choosing the Right Registry

Select an appropriate registry that aligns with the nature of the study and the research area.

6.4.2. Preparing the Protocol Document

Prepare a comprehensive protocol document that includes all necessary components.

6.4.3. Registering the Protocol

Register the protocol by submitting it to the chosen registry. Many registries have specific guidelines for registration.

6.4.4. Revisions and Updates

Update the protocol if any changes occur during the research process, ensuring transparency and accuracy.

> **Protocol registration** is a vital practice that ensures research transparency, promotes research integrity, and contributes to evidence-based decision-making. By registering research protocols in publicly accessible registries, researchers foster a culture of accountability and reliability in scientific research. As more researchers and stakeholders recognize the significance of protocol registration, it will continue to play an essential role in advancing knowledge and improving research practices in various fields.

6.5. Benefits of Protocol Registration

6.5.1. Research Integrity

Protocol registration promotes research integrity by minimizing biases, avoiding data dredging, and ensuring transparency in reporting.

6.5.2. Reproducibility and Replicability

Registered protocols facilitate reproducibility and replicability, enabling other researchers to verify and build upon existing research.

6.5.3. Credibility and Trustworthiness

Protocol registration enhances the credibility and trustworthiness of research findings, providing evidence that the study was pre-planned and conducted ethically.

6.5.4. Collaboration and Avoiding Redundancy

Registered protocols encourage collaboration among researchers and prevent duplication of efforts in the scientific community.

The Preferred Reporting Items for Systematic reviews and Meta-analyses for Protocols 2015 (PRISMA-P 2015). PRISMA-P consists of a 17-item checklist intended to facilitate the preparation and reporting of a robust protocol for the systematic review.

6.6. Challenges and Limitations

6.6.1. Time and Resource Constraints

Preparing a detailed protocol and registering it can be time-consuming and may require additional resources.

6.6.2. Variability in Registry Requirements

Different registries may have varying requirements and submission guidelines, causing confusion for researchers.

6.6.3. Language and Accessibility Barriers

Some registries may not support multiple languages, limiting the accessibility of research protocols.

6.7. Future Perspectives

6.7.1. Wider Adoption of Protocol Registration

Encouraging researchers and funding agencies to adopt protocol registration as a standard practice.

6.7.2. Improving Registry Infrastructure

Developing user-friendly registry platforms with standardized formats to simplify the registration process.

6.7.3. Promoting Research Transparency

Educating researchers about the importance of research transparency and the benefits of protocol registration.

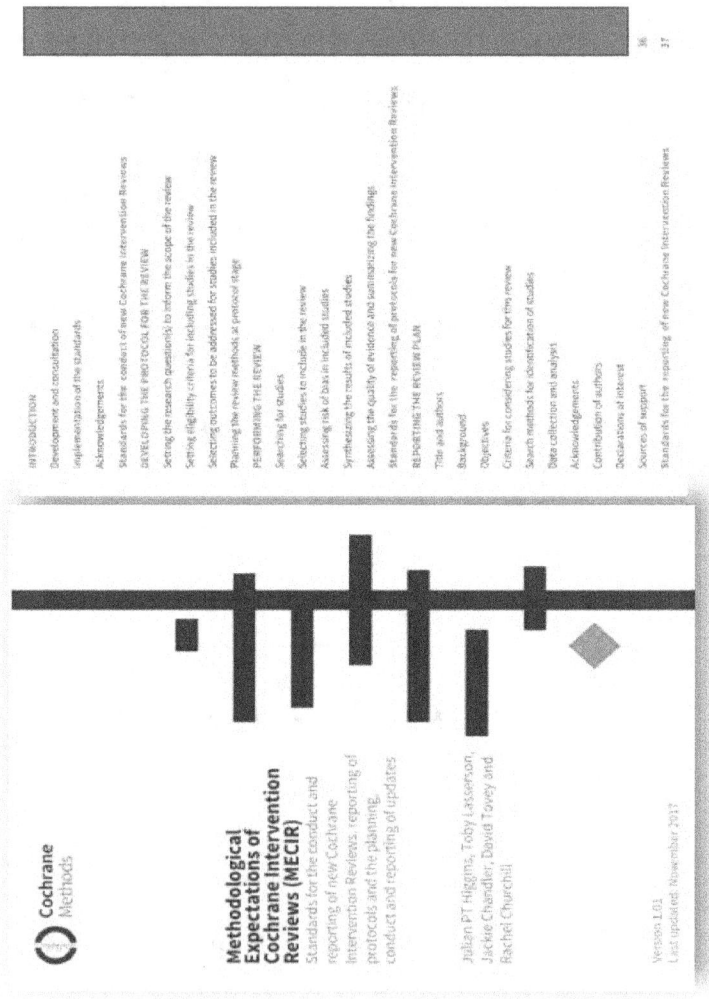

MECIR (Methodological Expectations for Cochrane Intervention Reviews) guidelines

MECIR (Methodological Expectations for Cochrane Intervention Reviews) is a set of guidelines developed by the Cochrane Collaboration for reporting protocols and conducting systematic reviews of interventions. MECIR outlines the methodological standards and expectations that Cochrane Review authors should adhere to when conducting and reporting intervention reviews. The MECIR Manual provides detailed guidance on the content and structure of systematic review protocols, ensuring consistency and transparency across Cochrane intervention reviews.

Here are some key components of the MECIR Manual for reporting protocols for Cochrane Intervention reviews:

Introduction: Introduce the systematic review, including the rationale for conducting the review and its relevance to clinical practice or policy.

Objectives: Clearly state the primary objective(s) of the systematic review, along with any secondary objectives.

Criteria for Considering Studies: Specify the eligibility criteria for including studies in the review, including the population, intervention, comparison, outcome (PICO) criteria, or other relevant criteria.

Search Methods: Describe the methods used to identify relevant studies, including the databases searched, search terms used, and any additional sources consulted (e.g., reference lists, trial registries).

Data Collection and Analysis: Outline the procedures for data extraction and analysis, including how data will be extracted from included studies, how data synthesis will be conducted, and any planned statistical methods or subgroup analyses.

Risk of Bias Assessment: Describe how the risk of bias will be assessed for included studies, including the criteria used and how discrepancies will be resolved.

Data Synthesis and Presentation: Explain how the findings from individual studies will be synthesized and presented, including the methods for combining data (e.g., meta-analysis) and presenting results (e.g., forest plots).

Additional Analyses: Specify any additional analyses planned, such as sensitivity analyses or subgroup analyses, and the rationale for conducting them.

Assessment of Heterogeneity: Describe how heterogeneity will be assessed and the methods for exploring and interpreting heterogeneity within the review.

Assessment of Reporting Biases: Outline any plans to assess reporting biases, such as publication bias, and the methods for addressing them.

Protocol Amendments: Describe any planned or completed amendments to the protocol, including the rationale for the changes.

Registration: Provide details of the systematic review protocol registration, including the registration number and registry name.

By following the MECIR Manual, Cochrane Review authors can ensure that their systematic review protocols are comprehensive, transparent, and methodologically sound, which enhances the quality and reliability of the resulting reviews.

PRISMA-P guideline

PRISMA-P (Preferred Reporting Items for Systematic Review and Meta-Analysis Protocols) is a guideline specifically designed for the reporting of protocols of systematic reviews and meta-analyses. It helps researchers ensure transparency and completeness in reporting key methodological details at the protocol stage, which is crucial for reducing bias and enhancing the reproducibility of systematic reviews.

Here are the key components of the PRISMA-P guideline:

Title: The title should clearly indicate that the document is a protocol for a systematic review or meta-analysis.

Background: Provide a brief introduction to the research question or topic, including its importance and relevance to clinical practice or policy.

Objectives: Clearly state the primary objective(s) of the systematic review or meta-analysis, along with any secondary objectives.

Methods: Describe the methods that will be used to conduct the systematic review or meta-analysis, including the following:

Eligibility criteria: Specify the population, intervention, comparison, outcome (PICO) criteria, or other inclusion/exclusion criteria.

Information sources: Describe the databases, registries, and other sources that will be searched for relevant studies.

Search strategy: Provide details of the search strategy, including search terms, Boolean operators, and any filters or limits applied.

Study selection: Outline the process for screening and selecting studies, including how discrepancies will be resolved.

Data extraction: Describe the data extraction process, including the variables to be extracted and how data will be recorded.

Data synthesis: Explain how data will be synthesized and analyzed, including any planned statistical methods or subgroup analyses.

Risk of bias assessment: Describe how the risk of bias will be assessed for included studies, if applicable.

Publication bias assessment: Outline any plans to assess publication bias, such as funnel plot analysis or other methods.

Data management: Describe how data will be managed, stored, and documented throughout the review process.

Registration: Provide details of the systematic review or meta-analysis protocol registration, including the registration number and registry name.

Funding: Disclose any sources of funding or support for the systematic review or meta-analysis.

Ethics approval: If applicable, indicate whether ethics approval is required or has been obtained for the review.

Protocol amendments: Describe any planned or completed amendments to the protocol, including the rationale for the 1

By following the PRISMA-P guideline, researchers can ensure that their systematic review or meta-analysis protocol is well-documented, transparent, and reproducible, which enhances the credibility and reliability of the research findings.

Table: Different study types with their reporting guidelines.

Study types	Guidelines for reporting
Randomised Trial	CONSORT
Observational studies	STROBE
Translational research	ARRIVE
Meta-analysis	PRISMA, PRISMA-NMA, PRISMA-IPD, MOSSE
Study protocol	PRISMA-P, SPIRIT, MECIR

Practice

Systematic Review Protocol for PROSPERO Registration

Hepatitis C Virus (HCV) Prevalence in India (given topic)

1. Introduction

1.1 Background Hepatitis C Virus (HCV) infection is a significant public health concern globally, and its prevalence varies across different regions. India, being a populous country with diverse demographics and risk factors, is likely to have varying HCV prevalence rates in different states and population groups.

This systematic review aims to synthesize the available evidence on HCV prevalence in India to provide a comprehensive overview of the burden of HCV infection in the country.

1.2 Objectives The primary objective of this systematic review is to estimate the overall prevalence of HCV infection in India. The secondary objectives include:

- Identifying regional variations in HCV prevalence within India.
- Analyzing HCV prevalence trends over time.
- Investigating the prevalence of HCV among specific population groups (e.g., high-risk populations, different age groups, rural vs. urban populations).

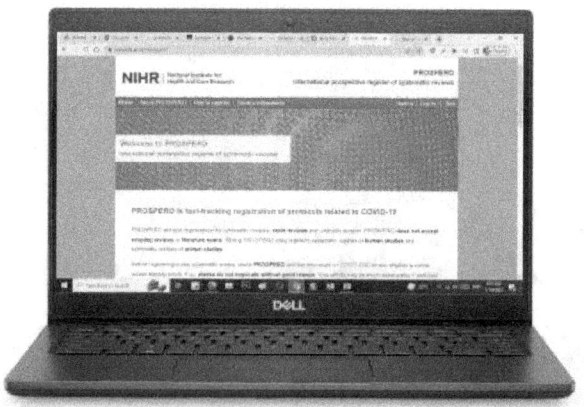

2. Methods

2.1 Eligibility Criteria

Population: Studies conducted on individuals residing in India, irrespective of age, gender, or risk factors, will be included.

Intervention/Exposure: Studies reporting the prevalence of HCV infection will be considered.

Comparator: There will be no comparator in this review.

Outcomes: The primary outcome will be the prevalence of HCV infection in India. Secondary outcomes may include regional prevalence, temporal trends, and subgroup prevalence.

Study Designs: Observational studies (cross-sectional, cohort, and case-control) that report original data on HCV prevalence in India will be eligible for inclusion.

2.2 Information Sources

Electronic Databases: PubMed/MEDLINE, Embase, Scopus, Web of Science, and IndMED will be searched for relevant studies.

Grey Literature: Grey literature sources, such as conference abstracts, theses, and reports, will also be searched for additional studies.

Reference Lists: The reference lists of included studies and relevant systematic reviews will be hand-searched for additional citations.

Experts: Experts in the field will be consulted to identify any unpublished or ongoing studies.

2.3 Search Strategy

The search strategy will be developed using a combination of Medical Subject Headings (MeSH) terms and relevant keywords related to HCV prevalence and India. The search strategy will be adapted for each database. The full search strategy will be reported in the final systematic review report.

2.4 Study Selection

Two independent reviewers will screen titles and abstracts of retrieved records against the eligibility criteria. Full-text articles of potentially eligible studies will be assessed independently by the same reviewers. Disagreements will be resolved through discussion or by consulting a third reviewer.

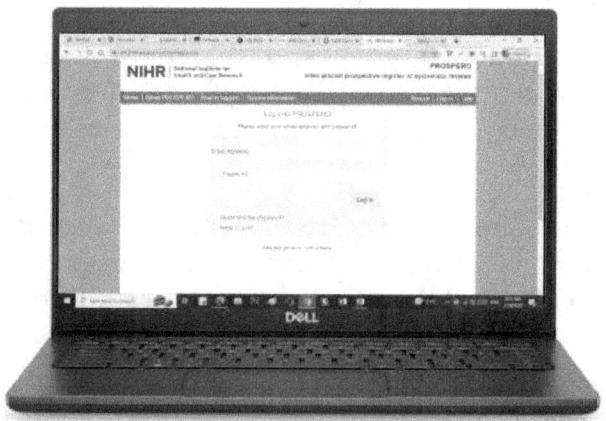

2.5 Data Extraction

A standardized data extraction form will be developed and pilot-tested. Two independent reviewers will extract data from included studies, including study characteristics (author, publication year, study design), participant demographics, HCV prevalence data, and any relevant subgroup information. Any discrepancies will be resolved through discussion or by involving a third reviewer.

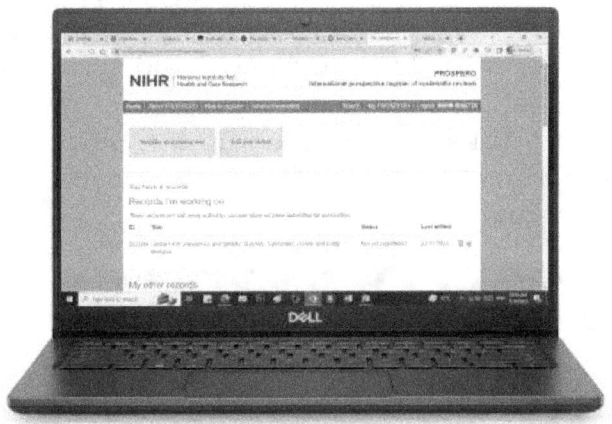

2.6 Risk of Bias Assessment

The risk of bias in individual studies will be assessed using appropriate tools (e.g., Newcastle-Ottawa Scale for observational studies). Two independent reviewers will assess the risk of bias, and any disagreements will be resolved through discussion or by consulting a third reviewer.

2.7 Data Synthesis

A narrative synthesis of the findings will be conducted, presenting the overall prevalence of HCV in India and exploring regional variations and subgroup prevalence. If data permit, a meta-analysis will be performed using a random-effects model. Heterogeneity will be assessed using the I^2 statistic.

2.8 Reporting

This systematic review protocol will be registered in the International Prospective Register of Systematic Reviews (PROSPERO) to ensure transparency and prevent duplication. The final systematic review report will adhere to the PRISMA (Preferred Reporting Items for Systematic Reviews and Meta-Analyses) guidelines.

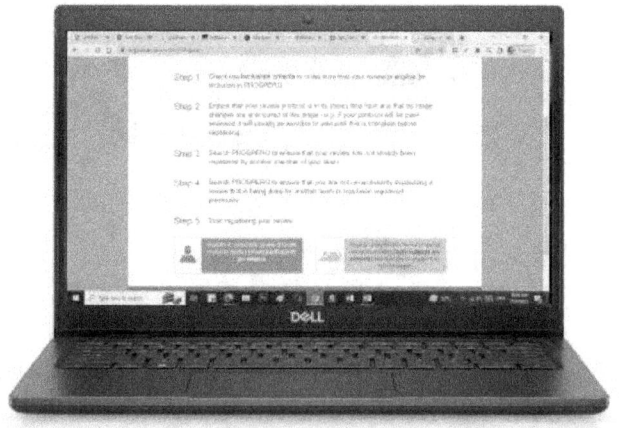

3. Ethics and Dissemination

As this systematic review will use publicly available data, ethical approval is not required. The findings of this review will be disseminated through a peer-reviewed publication and presented at relevant conferences to contribute to the evidence base on HCV prevalence in India and inform public health policies and interventions.

4. Amendments

Any modifications to this protocol during the conduct of the systematic review will be documented and explained in the final systematic review report.

Synthesizing Evidence: The Art of Systematic Review

These screening questions check whether your review is eligible for inclusion in PROSPERO and avoid wasting your time if it is not eligible.

Will your registration record be in English?
YES

Is this a scoping, literature or mapping review?
NO

Does your review include a health outcome with direct relevance to human health? (e.g. reviews of educational interventions to improve maths skills are not eligible, reviews of educational interventions to promote breastfeeding are eligible)
YES

Is this a Cochrane review?
NO

Is this a mini or partial review done for a training course or classwork or are you using the system to learn how to register?
NO

PROSPERO does not have resources to process applications for reviews done only for training purposes. This includes mini reviews restricted to a subset of eligible studies, demonstrator reviews where a whole class does the same review, or any other projects that are less than full systematic reviews.
For learning purposes you may download and complete the PROSPERO form as a PDF document. If you do complete the form online, please save this in your own space and do not SUBMIT it for publication.
YES

Have you written a protocol?
PROSPERO registration captures key information about the design and conduct of a planned systematic review. It is not a full protocol. We strongly encourage you to write a full protocol before completing the PROSPERO registration form (although you may proceed without doing this).
YES NO

Will more than one person be involved in the systematic review?
We strongly recommend that you follow best practice and include more than one person in the review team. PROSPERO will not accept registrations unless there is more than one person conducting the review. You must include details of the other author(s) in the registration form.
YES

Do you intend to publish the results of your systematic review and/or make them publicly available when completed?
PROSPERO aims to increase transparency and help prevent unintended duplication of effort. This requires that the results of systematic reviews should be made publicly available e.g. by publication in an academic journal, posting in a research repository or being made available on a permanent website. We therefore do not accept registrations from systematic reviews that will not be made available to others e.g. projects that are internal to an organization or company, or masters dissertations if it is known that these will not be shared.
YES

Have you started your review?
NO

Please now go ahead and register your review.

Donald Thomas Campbell

Donald Thomas Campbell (November 20, 1916 – May 6, 1996) was an American social scientist. He is noted for his work in methodology. He coined the term evolutionary epistemology and developed a selectionist theory of human creativity. A Review of General Psychology survey, published in 2002, ranked Campbell as the 33rd most cited psychologist of the 20th century. In June 1981, working with Alexander Rosenberg, Campbell organized an international conference held at Cazanovia, New York, to formulate the program of what he called an "Epistemologically Relevant Sociology of Science" (ERRES). By Campbell's own account, this project was at least premature. Campbell was elected to the

American Academy of Arts and Sciences and the National Academy of Sciences in 1973. In 1975, Campbell served as president of the American Psychological Association. He was elected to the American Philosophical Society in 1993.

Campbell also had a vision for how public policy could be improved through use of experimentation. He argued for a more collaborative method of public policy that involved various stakeholders and that used experimentation and data as a guide for decision making. The vision of this was laid out in his essay, "The Experimenting Society". His book *Experimental and Quasi-Experimental Designs for Research* became the standard in policy-evaluation circles. Campbell did not start out intending to be a program evaluator, but his devotion to understanding causality, human behaviour, and how to solve social questions led him there. In the 1990s, Campbell's formulation of the mechanism of "blind-variation-and-selective-retention" (BVSR) was further developed and extended to other domains under the labels of "universal selection theory" or "universal selectionism" by Gary Cziko, Mark Bickhard, and Francis Heylighen.

CHAPTER-7

PROBLEM FORMULATION, INCLUSION AND EXCLUSION CRITERIA, AND SETTING THE SCOPE IN RESEARCH

In the research process, problem formulation, inclusion and exclusion criteria, and setting the scope are crucial steps in defining the focus and boundaries of a study. These components play a pivotal role in shaping the research question, identifying the target population, and determining which studies or data will be included in the final analysis. This comprehensive guide explores the importance of problem formulation, the establishment of inclusion and exclusion criteria, and the process of setting the scope in research.

We will delve into the key considerations, methods, and strategies involved in each step to ensure a systematic and robust approach.

7.1. Problem Formulation

7.1.1. Definition and Purpose

Problem formulation is the process of identifying and defining a clear, concise, and well-structured research question or problem that a study aims to address. It involves identifying the key issues or gaps in knowledge, exploring the relevant literature, and formulating a research question that guides the entire research process.

7.1.2. Importance of Problem Formulation

Focus and Clarity: Problem formulation helps researchers maintain a clear focus on the research question and ensures that the study addresses a specific issue or problem.

Relevance and Significance: A well-formulated research question highlights the relevance and significance of the study, making it more appealing to stakeholders, funders, and readers.

Feasibility: A well-defined research question facilitates the design of a feasible study,11 as it guides the choice of study design, data collection methods, and data analysis techniques.

Guidance for Data Collection and Analysis: Problem formulation provides guidance for selecting appropriate data sources and helps determine the data analysis methods that align with the research question.

7.1.3. Process of Problem Formulation

Identify the Research Area: The first step is to identify the general area of research interest. This could be a specific field, topic, or problem that the researcher is interested in exploring.

Review Existing Literature: Conduct a thorough literature review to identify gaps in knowledge, unanswered questions, or controversies within the chosen research area.

Narrow Down the Focus: Based on the literature review, narrow down the focus to a specific research question that is manageable, relevant, and feasible to address.

Use the PICO Framework: The PICO framework (Population, Intervention, Comparison, Outcome) is a helpful tool for formulating research questions in healthcare and clinical research. It helps structure the question by defining the target population, the intervention or exposure of interest, the comparison group (if applicable), and the desired outcome.

Be Specific and Measurable: Ensure that the research question is specific, measurable, and feasible to answer. Avoid broad or vague questions that may be difficult to address in a single study.

7.2. Inclusion and Exclusion Criteria

7.2.1. Definition and Purpose

Inclusion and exclusion criteria are predetermined criteria that researchers use to select and filter studies or data for inclusion in a research study or systematic review. These criteria define the characteristics and attributes that a study must possess to be considered relevant to the research question.

7.2.1.1. Importance of Inclusion and Exclusion Criteria

Consistency: Inclusion and exclusion criteria ensure that the study selection process is consistent and objective, reducing the potential for bias in study selection.

Relevance: Inclusion and exclusion criteria help ensure that only studies that are directly relevant to the research question are included in the analysis.

Study Validity: By applying strict criteria, researchers can enhance the validity and reliability of the study findings.

Manageability: Inclusion and exclusion criteria help manage the scope of the study, as they define the boundaries of the research.

7.2.2. Process of Defining Inclusion and Exclusion Criteria

Align with Research Question: The criteria should align with the research question and the specific objectives of the study.

Population: Define the characteristics of the target population that will be included in the study. This may include age, gender, health status, and other relevant demographic factors.

Study Design: Specify the types of study designs that are eligible for inclusion (e.g., randomized controlled trials, observational studies).

Intervention and Exposure: Describe the specific intervention or exposure that the studies must have to be considered relevant.

Outcomes: Identify the primary and secondary outcomes of interest that the studies must report on.

Language and Publication Status: Decide whether language restrictions and publication status (published or unpublished) will be applied.

Time Frame: Set a time frame for the publication date of the studies to be included.

7.3. Setting the Scope

7.3.1. Definition and Purpose

Setting the scope of a research study involves defining the boundaries and extent of the investigation. It encompasses the depth and breadth of the research, the study population, the time frame, the geographic location, and other relevant parameters that frame the research context.

7.3.2. Importance of Setting the Scope

Focus and Clarity: Setting the scope provides focus and clarity to the research, preventing it from becoming overly broad or unmanageable.

> **Problem formulation, inclusion and exclusion criteria, and setting the scope** are crucial steps in the research process that lay the foundation for a successful study. A well-formulated research question guides the entire research process and ensures a clear focus on the specific issue being addressed. Inclusion and exclusion criteria ensure that the study selects relevant and appropriate studies or data for analysis, while setting the scope defines the boundaries and parameters of the research investigation. By carefully considering these components and following a systematic approach, researchers can conduct studies that are well-structured, focused, and relevant, leading to meaningful and impactful research outcomes.

Manageability: A well-defined scope ensures that the research project is feasible and can be completed within the available resources and time frame.

Relevance: By setting the scope, researchers ensure that the study remains relevant to the specific research question and objectives.

Consistency: A defined scope helps maintain consistency and uniformity throughout the research process.

7.3.3. Process of Setting the Scope

Define the Research Area: Clearly articulate the general research area or topic that the study will address.

Specify the Research Question: Ensure that the research question is specific and well-defined to provide a clear scope.

Identify the Study Population: Define the target population that the study will focus on, including any relevant characteristics or demographic factors.

Geographic Location: If applicable, specify the geographic location or context of the study.

Time Frame: Set the time frame for the study, indicating the start and end dates of the data collection or observation period.

Data Sources: Identify the sources of data that will be used for the study, such as databases, surveys, or interviews.

Boundaries: Clearly define the boundaries of the research by specifying any restrictions or exclusions related to the research question.

David Lawrence Sackett

David Lawrence Sackett (November 17, 1934 – May 13, 2015) was an American-Canadian physician and a pioneer in evidence-based medicine. He is known as one of the fathers of Evidence-Based Medicine. He founded the first department of clinical epidemiology in Canada at McMaster University, and the Oxford Centre for Evidence-Based Medicine. He is well known for his textbooks Clinical Epidemiology and Evidence-Based Medicine. One of his more famous quotes is: "Half of what you learn in medical school is dead wrong."

His contributions to research methodology included ways to detect and reduce bias in clinical research, and ways to design, conduct, and report randomized clinical trials. David Sackett is widely regarded as one of 3 "fathers" of modern clinical epidemiology (along with Archie Cochrane of the UK and Alvan Feinstein of the USA). Clinical epidemiology is a research discipline based on the methods of epidemiology (and other scientific pursuits, notably biostatistics, the behavioral sciences, and health economics), applied to understanding the nature of health care problems and, especially, their management. Thus, it is a bridging discipline, linking research to clinical practice. Typical topics include the cause, diagnosis, course (prognosis, clinical prediction), prevention, treatment, and amelioration of health disorders, and the improvement and cost-effectiveness of health services.

Sackett was the founding chair of the first department of Clinical Epidemiology and Biostatistics in the world at McMaster University in Hamilton, Ontario, in 1967, and extensively contributed to the development of research methods through his writings

CHAPTER-8

FORMATION OF REVIEW QUESTIONS FOR SYSTEMATIC REVIEW

A systematic review is a rigorous and transparent approach to evidence synthesis that aims to answer specific research questions by systematically identifying, appraising, and synthesizing the available evidence. The formation of well-defined and relevant review questions is a critical step in the systematic review process. Review questions guide the entire review, from the development of the protocol to the interpretation and dissemination of findings. This chapter provides a comprehensive guide to the formation of review questions for systematic reviews. It explores the key principles, considerations, and methods involved in crafting focused and meaningful review questions.

8.1 Understanding the Importance of Review Questions

The Role of Review Questions

Review questions serve as the backbone of a systematic review, defining the scope and objectives of the study. They guide the search for relevant studies, the selection of inclusion and exclusion criteria, data extraction, data synthesis, and the interpretation of results.

Specificity and Clarity

Review questions should be specific and clear, leaving no room for ambiguity. A well-defined question helps researchers stay focused on the research objectives, ensuring that the review produces meaningful and actionable results.

Relevance to Stakeholders

Review questions should address relevant and important topics that have practical implications for stakeholders, such as healthcare providers, policymakers, or patients.

8.2 The Formulation Process

Defining the Research Area

Identify the general area of interest for the systematic review. This may be a specific field, topic, or problem that requires evidence synthesis.

Conducting a Scoping Review

Before finalizing the review question, conduct a scoping review to explore the existing literature, identify gaps, and assess the feasibility of the research question.

Engaging Stakeholders

Engage relevant stakeholders, such as policymakers or practitioners, to ensure that the review question addresses their needs and concerns.

Using the PICO Framework

The PICO (Population, Intervention, Comparison, Outcome) framework is a valuable tool for formulating review questions, particularly in healthcare and clinical research. It structures the question by defining the target Population, the Intervention or exposure of interest, the Comparison group (if applicable), and the Outcome of interest.

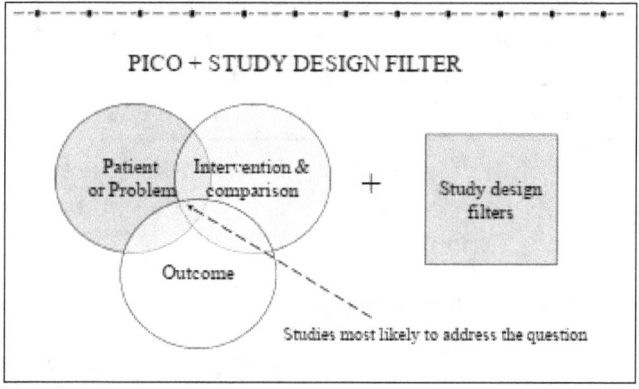

8.3. Asking the Right Questions

Different Types of Review Questions

Effectiveness Questions: These questions assess the effectiveness of a specific intervention or treatment compared to a control or alternative intervention.

Diagnostic Questions: Diagnostic questions assess the accuracy of diagnostic tests or tools in identifying a particular condition or disease.

Prognostic Questions: Prognostic questions examine the factors or variables that predict the outcome of interest.

> **The formation of review questions** is a crucial step in the systematic review process, providing the foundation for the entire study. Well-defined and relevant review questions ensure that the review remains focused, transparent, and meaningful, leading to evidence-based decision-making and improved practice in various fields. By following the principles and strategies outlined in this chapter, researchers can formulate review questions that address important issues, engage stakeholders, and contribute to advancing knowledge in their respective fields.

Economic Questions: Economic questions assess the cost-effectiveness or cost-benefit of an intervention or policy.

Framing the Questions

Background Questions: Background questions provide general information on a topic, such as prevalence or risk factors, to provide context for the review.

Foreground Questions: Foreground questions are specific and focused on interventions, comparisons, outcomes, and study populations, guiding the main analysis.

Using the PICOS Framework

The PICOS (Population, Intervention, Comparison, Outcome, Study design) framework is another approach to formulating review questions, particularly in healthcare and clinical research.

8.4. Addressing Sub-Questions

Review questions may sometimes require breaking down into sub-questions to explore different aspects of the main question. Sub-questions can help organize the review process and provide a more comprehensive understanding of the topic.

8.5. Writing and Refining the Review Questions

Keeping the Questions Clear and Concise

Ensure that the review questions are written in a clear and concise manner,

avoiding unnecessary jargon or technical language.

Peer Review and Expert Input

Seek feedback from colleagues and experts in the field to refine the review questions and ensure they are well-structured and relevant.

Examples of Well-Formulated Review Questions

1: Effectiveness Question

"Among adults with type 2 diabetes, does aerobic exercise compared to no exercise reduce HbA1c levels?"

2: Diagnostic Question

"In children with suspected appendicitis, what is the accuracy of ultrasound compared to computed tomography (CT) for diagnosing appendicitis?"

3: Prognostic Question

"What factors predict long-term outcomes (e.g., mortality, disability) in patients with traumatic brain injury?"

4: Economic Question

"What is the cost-effectiveness of vaccination programs for preventing influenza in the elderly population?"

Review Question

Population

Disease or condition, Stage, Severity, Demographic characteristics (age, gender, etc.)

Intervention:

Type of intervention or exposure, Dose, duration, timing, route, etc.

Comparison:

Absence of risk or treatment Placebo or alternative therapy

Outcome:

Risk or protective, Dichotomous or continuous, Type: mortality, morbidity, quality of life, etc.

CHAPTER-9

SYSTEMATIC METHODS OF LITERATURE SEARCH

Literature search, the systematic exploration of published and unpublished materials, is a crucial process that enables researchers to access relevant information, identify gaps in knowledge, and provide a solid foundation for their work. This chapter explores the significance of literature search in research, the key components involved, and the methods to conduct a comprehensive and effective literature search.

9.1. Understanding Literature Search

9.1.1. Definition and Purpose

Literature search is the process of systematically and comprehensively exploring various sources to identify relevant research studies, articles, books, reports, and other published materials related to a specific research question or topic. The primary purpose of literature search is to gather information and evidence to support a research study, review, or any other academic inquiry.

9.1.2. Role in Evidence-Based Practice

Literature search plays a vital role in evidence-based practice by providing clinicians, practitioners, policymakers, and researchers with up-to-date and reliable evidence to inform decision-making.

9.2. Importance of Literature Search in Research

9.2.1. Identifying Research Gaps

A comprehensive literature search helps identify gaps in existing knowledge, thereby guiding researchers to address important research questions.

9.2.2. Literature Review and Synthesis

A literature search serves as the foundation for literature reviews and systematic reviews, ensuring that all relevant studies are included in the analysis.

9.2.3. Validating Research Ideas

Literature search allows researchers to determine the feasibility and novelty of their research ideas by assessing the availability of relevant literature.

9.2.4. Supporting Hypotheses and Arguments

Literature search provides evidence to support or refute hypotheses and arguments, enhancing the credibility of research findings.

9.2.5. Identifying Best Practices

For practitioners and policymakers, literature search helps identify best practices and evidence-based interventions in various fields.

9.3. Key Components of Literature Search

9.3.1. Research Question or Topic

Clearly define the research question or topic to guide the literature search process effectively.

9.3.2. Keywords and Search Terms

Identify relevant keywords and search terms to use in databases and search engines for retrieving relevant literature.

9.3.3. Search Strategies

Develop a systematic search strategy that includes both broad and specific terms to ensure comprehensive coverage of the literature.

9.3.4. Selection of Databases: Choose appropriate databases and search engines

based on the research topic and discipline to access relevant literature.

9.3.5. Inclusion and Exclusion Criteria

Set clear inclusion and exclusion criteria to filter and select relevant studies during the screening process.

9.3.6. Filters and Limiters

Use filters and limiters, such as publication date, language, study design, or geographical location, to refine the search results.

9.4. Methods for Conducting a Literature Search

9.4.1. Database Searching

Utilize academic databases such as PubMed, Scopus, Web of Science, or Google Scholar to access scholarly articles and research studies.

9.4.2. Hand Searching

Supplement database searches with hand searching of journals, conference proceedings, and reference lists of relevant articles.

9.4.3. Grey Literature

Include grey literature, such as unpublished reports, theses, dissertations, and conference abstracts, to capture additional relevant information.

Literature search is a fundamental process that underpins the success of any research endeavour. It plays a pivotal role in identifying existing knowledge, informing research questions, and supporting evidence-based practice. A well-executed literature search helps researchers access relevant and reliable information, validates research ideas, and enhances the rigor and credibility of their work. By employing effective search strategies, critically appraising search results, and staying updated on best practices, researchers can harness the full potential of literature search to drive scientific advancement and evidence-based decision-making in diverse fields of research.

9.4.4. Contacting Experts

Contact experts in the field for recommendations on relevant literature and unpublished research.

9.5. Evaluating the Quality of Search Results

9.5.1. Assessing Relevance

Screen the search results based on the inclusion and exclusion criteria to identify relevant studies.

9.5.2. Appraising Source Credibility

Evaluate the credibility and reliability of the sources by assessing the authorship, affiliation, peer-review process, and publication venue.

9.6. Data Management and Organization

9.6.1. Reference Management Software

Use reference management software, such as EndNote or Zotero, to organize and manage the retrieved references.

9.6.2. Data Extraction

Systematically extract relevant data from the selected studies to facilitate data synthesis and analysis.

9.7. The Challenges of Literature Search

9.7.1. Information Overload

The vast amount of available literature can make it challenging to identify the most relevant and important studies.

9.7.2. Language Barriers

Language restrictions can limit access to literature published in languages other than English.

9.7.3. Publication Bias

Publication bias can lead to an overrepresentation of positive or significant findings, potentially skewing the evidence available.

9.8. Enhancing the Rigor of Literature Search

9.8.1. Peer Review

Seek feedback from colleagues or experts in the field to validate the search strategy and findings.

9.8.2. Reproducibility and Documentation

Document the search strategy and results comprehensively to ensure reproducibility and transparency.

CHAPTER-10

SEARCH TERMS, SOURCES OF SEARCH, AND E-DATABASES FOR SYSTEMATIC REVIEW

The success of a systematic review heavily relies on the selection of appropriate search terms, sources of search, and electronic databases. This chapter aims to provide a comprehensive guide to selecting relevant search terms, identifying diverse sources of search, and leveraging electronic databases effectively to conduct a robust systematic review.

10.1. Selecting Relevant Search Terms

10.1.1. Defining the Research Question

A clear and well-defined research question serves as the foundation for selecting relevant search terms. The research question should be specific, focused, and aligned with the systematic review's objectives.

10.1.2. Identifying Keywords and Synonyms

Identify relevant keywords and synonyms that capture the main concepts of the research ques10tion. Thesauruses and controlled vocabularies can assist in identifying appropriate synonyms.

10.1.3. Using Boolean Operators

Utilize Boolean operators (AND, OR, NOT) to combine search terms logically and refine search results.

10.1.4. Truncation and Wildcards

Applying truncation and wildcards allows for variations of search terms, enhancing the search's sensitivity and comprehensiveness.

10.1.5. MeSH Terms and Indexing

For medical and healthcare topics, utilize Medical Subject Headings (MeSH) terms to leverage the indexing of research articles in databases like PubMed.

10.2. Sources of Search for Systematic Review

10.2.1. Academic Databases

Explore academic databases like PubMed, Scopus, Web of Science, and Embase to access a wide range of scholarly articles and research studies.

10.2.2. Grey Literature

Incorporate grey literature sources, including conference abstracts, theses, dissertations, and reports, to capture unpublished or difficult-to-find research.

10.2.3. Institutional Repositories

Check institutional repositories of universities and research organizations for the availability of full-text articles and reports.

10.2.4. Preprint Servers

For cutting-edge research, search preprint servers like MedRixv, arXiv and bioRxiv, where researchers share preliminary findings before peer review.

10.2.5. Government Websites and Registries

Government websites, clinical trial registries, and public health agencies often host valuable reports and research publications.

10.2.6. Professional Organizations

Explore the websites and publications of professional organizations related to the research topic.

10.3. Electronic Databases for Systematic Review

10.3.1. PubMed

PubMed is a widely used database for biomedical and life sciences literature. It offers advanced search features and MeSH indexing.

10.3.2. Scopus

Scopus covers a broad range of disciplines, providing access to peer-reviewed articles and conference proceedings.

10.3.3. Web of Science

Web of Science offers a multidisciplinary collection of scholarly articles and citation databases, enabling citation analysis.

10.3.4. Embase

Embase specializes in biomedical literature and offers extensive drug and pharmaceutical coverage.

10.3.5. PsycINFO

PsycINFO is a valuable resource for psychology and behavioral science research.

10.3.6. CINAHL

CINAHL is a specialized database for nursing and allied health research.

10.4. Constructing Effective Search Strategies

10.4.1. Study Design Filters

Utilize study design filters to focus on specific types of research studies (e.g., randomized controlled trials, observational studies).

10.4.2. Language and Time Filters

Apply language and time filters to narrow down the search results.

10.4.3. Exploring Boolean and Nesting

Employing complex Boolean search strategies and nesting ensures comprehensive and precise retrieval of relevant literature.

10.4.4. Search Query Documentation

Documenting search strategies in detail enhances transparency and reproducibility.

10.5. Managing Search Results

10.5.1. Reference Management Software

Use reference management software, such as EndNote or Zotero, to organize and manage the retrieved references.

10.5.2. Screening and Selection

Systematically screen and select studies based on inclusion and exclusion criteria.

10.5.3. Data Extraction

Extract relevant data from selected studies using predefined data extraction forms.

10.6. Challenges and Limitations

10.6.1. Language Barriers

Language restrictions may limit access to literature published in languages other than English.

10.6.2. Publication Bias

Publication bias can skew the evidence available, leading to an overrepresentation of positive findings.

10.6.3. Time and Resource Constraints

Conducting a comprehensive literature search can be time-consuming and resource-intensive.

10.7. Quality Assessment of Search and Reporting

10.7.1. PRISMA

Adhering to PRISMA (Preferred Reporting Items for Systematic Reviews and Meta-Analyses) guidelines enhances the transparency and quality of reporting the literature search process.

10.7.2. Peer Review and Validation

Seek peer review and validation of the search strategies to improve the quality and comprehensiveness of the literature search.

Selecting relevant search terms, identifying diverse sources of search, and effectively using electronic databases are critical aspects of conducting a successful systematic review. By following the principles and strategies outlined in this chapter, researchers can conduct comprehensive and rigorous literature searches that serve as the foundation for evidence-based practice, informed decision-making, and advancing knowledge in their respective fields.

10.8. Search strategies

In a systematic review, search terms are specific keywords and phrases used to identify relevant studies and literature related to the research question or topic under investigation. These search terms help researchers conduct a comprehensive and focused search across various databases and sources to gather all relevant evidence for the systematic review. Here are different types of search terms commonly used in a systematic review:

Main Concepts: These are the key elements of the research question or topic. Main concepts form the core of the search terms and represent the major themes or variables being studied. For example:

Research Question: "Does physical exercise reduce symptoms of depression in adults?"

Main Concepts: "Physical exercise," "symptoms of depression," "adults"

Keywords: Keywords are specific words or phrases directly related to the research question. These are the terms that researchers would expect to find in the titles, abstracts, or full texts of relevant articles. For example:

Keywords: "exercise," "physical activity," "depression," "mood disorder," "adult population"

Synonyms and Alternate Terms: Researchers should consider using synonyms and alternate terms for the main concepts and keywords to ensure comprehensive coverage. Different studies may use different terminology to refer to the same concept. For example:

Synonyms: "workout," "training," "aerobic activity," "affective disorder," "mental health," "mood disturbance," "grown-ups"

MeSH Terms: Medical Subject Headings (MeSH) are standardized vocabulary terms used in the PubMed database and some other biomedical databases. Using MeSH terms allows for more precise and consistent retrieval of relevant studies. For example:

MeSH Terms: "Exercise," "Depression," "Adult"

Boolean Operators: Boolean operators (AND, OR, NOT) are used to combine search terms logically and refine search results. For example:

Search Query: ("Exercise" OR "Physical activity" OR "Workout") AND ("Depression" OR "Mood disorder" OR "Affective disorder") AND "Adult"

Truncation and Wildcards: Truncation and wildcards allow for variations of search terms and can help capture different forms of the same word. For example:

Truncation: "Depress*" (captures "depression," "depressive," "depressed," etc.)

Wildcard: "Exerci?" (captures "exercise" or "exercising")

Study Design Filters: In some systematic reviews, researchers may include specific study design terms in their search to focus on certain types of research. For example:

Study Design Filter: "randomized controlled trial," "observational study," "cohort study"

Publication Date and Language Filters: Researchers may use filters to limit the search results to a specific publication date range or language. For example:

Publication Date Filter: "Publication date: 2010-2023"

Language Filter: "English language only"

Geographic Filters: If the systematic review is focused on a particular geographic region, researchers may include geographic terms as part of the search. For example:

Geographic Filter: "USA," "Europe," "Asia"

Remember, the choice and combination of search terms will depend on the specific research question, the scope of the systematic review, and the databases being used. A well-designed search strategy with carefully selected search terms is crucial for ensuring that the systematic review captures all relevant evidence and provides a comprehensive and unbiased synthesis of the available literature.

Flow Chart: Comprehensive Search for a Systematic Review on a Given Topic

1. Define Research Question and Scope

a. Clearly define the research question to guide the search process.
b. Determine the scope of the systematic review, including inclusion and exclusion criteria.

2. Identify Relevant Keywords and Concepts

a. Identify key concepts and keywords related to the research question.
b. Include synonyms and alternate terms to ensure comprehensive coverage.

3. Develop a Search Strategy

a. Create a structured search string using Boolean operators (AND, OR, NOT) to combine keywords logically.
b. Utilize truncation and wildcards to capture variations of search terms.

4. Select Databases and Sources

a. Choose appropriate databases based on the research topic and discipline (e.g., PubMed, Scopus, Web of Science).
b. Include grey literature sources, such as conference abstracts, theses, and reports.

5. Execute the Search

a. Execute the search using the developed search strategy in each selected database and source.
b. Record the date of each search and save the search results.

6. Remove Duplicates

a. Remove duplicate records using reference management software (e.g., EndNote, Zotero).

7. Title and Abstract Screening

a. Screen the titles and abstracts of retrieved records against inclusion and exclusion criteria.
b. Independently assess each study's relevance by two or more reviewers.
c. Resolve discrepancies through discussion or involving a third reviewer if necessary.

8. Full-Text Screening

a. Retrieve full-text articles for potentially eligible studies identified during title and abstract screening.
b. Conduct a detailed assessment of each full-text article against inclusion and exclusion criteria.
c. Document reasons for excluding studies during full-text screening.

9. Grey Literature Search

a. Perform a separate search for grey literature using appropriate sources and search terms.
b. Include grey literature records in the screening process.

Iain Chalmers

Iain Geoffrey Chalmers (born 3 June 1943) is a British health services researcher, one of the founders of the Cochrane Collaboration, and coordinator of the James Lind Initiative, which includes the James Lind Library and James Lind Alliance. Between 1978 and 1992, he was the first director of the National Perinatal Epidemiology Unit in Oxford. There, Chalmers led the development of the electronic Oxford Database of Perinatal Trials (ODPT) and a collection of systematic reviews of randomized trials of care in pregnancy and Childbirth. In 1992, Chalmers was appointed director of the UK Cochrane Centre, leading to the development of the international Cochrane Collaboration.

Subsequently, he became founding editor of the James Lind Library, which documents the history and evolution of fair trials of treatments, and helped to establish the James

Lind Alliance, a non-profit organization that "aims to identify the most important gaps in knowledge about the effects of treatments". Throughout his career, Chalmers has been a staunch advocate for evidence-based medicine (EBM), emphasizing the importance of basing clinical decisions on the best available evidence from scientific research. He has promoted the rigorous evaluation of medical interventions through randomized controlled trials (RCTs) and systematic reviews, advocating for transparency, accountability, and patient-centred care.

Chalmers has also been actively involved in promoting research ethics and integrity. He co-founded the James Lind Initiative, which aims to raise awareness of fair tests of treatments in healthcare and improve the quality of research. He has been vocal about the need for ethical conduct in clinical trials and research studies, emphasizing the importance of protecting the rights and welfare of research participants.

His contributions to evidence-based medicine and healthcare research have been widely recognized. His legacy extends far beyond his individual achievements. His pioneering efforts in promoting evidence-based medicine and establishing the Cochrane Collaboration have had a profound impact on medical practice, research methodology, and healthcare policy worldwide.

CHAPTER-11

DOCUMENTATION AND REPORTING OF SEARCHES

A crucial aspect of conducting a systematic review is the documentation and reporting of the search process. Proper documentation ensures the reproducibility of the review, while comprehensive reporting allows readers to understand the search strategy's scope and limitations. This chapter provides a comprehensive guide to the documentation and reporting of searches for systematic reviews, outlining the key components, best practices, and the impact of transparent reporting on the credibility and reliability of the review.

11.1. Importance of Documentation and Reporting

11.1. Reproducibility

Clear and detailed documentation allows other researchers to replicate the search strategy and verify the results independently.

11.2. Transparency

Transparent reporting provides readers with insights into the methods used, enhancing the credibility and trustworthiness of the systematic review.

11.3. Minimizing Bias

Comprehensive documentation minimizes the risk of bias in the search process, ensuring that all relevant studies are included.

11.4. Facilitating Peer Review

Complete reporting enables peer reviewers to assess the quality and validity of the search strategy and results.

11.2. Components of Documentation

11.2.1. Search Protocol

Develop a search protocol outlining the entire search strategy, including search terms, databases, and any filters or limits applied.

11.2.2. Database Information

Record the names and URLs of the databases searched, along with the date of the search.

11.2.3. Search Strings and Queries

Document the complete search strings and queries used for each database, including the combination of keywords, Boolean operators, and truncation.

> The documentation and reporting of searches are critical components of a systematic review, enhancing the transparency, credibility, and reliability of the review process. By adhering to best practices and ensuring comprehensive reporting, researchers can foster a culture of accountability and reproducibility, advancing evidence-based practice and informed decision-making in various fields of research. A systematic and well-documented search strategy lays the groundwork for a robust systematic review, enabling researchers to synthesize the available evidence effectively and contribute to the advancement of knowledge in their respective disciplines.

2.4. Search Filters and Limits

Record any filters or limits applied, such as publication date, language, study design, or geographical location.

11.2.5. Grey Literature Sources

Include a list of sources searched for grey literature, such as conference abstracts, theses, and reports.

11.2.6. Correspondence with Experts

Document any correspondence with experts or authors to obtain additional studies or information.

11.2.7. Search Iterations

If the search strategy was refined or updated during the review process, document the different iterations of the search.

11.3. Best Practices for Documentation

11.3.1. Version Control

Maintain version control for the search protocol, ensuring that any changes or updates are clearly documented.

11.3.2. Timestamps

Include timestamps for significant steps in the search process, such as the date of search execution and the date of data extraction.

11.3.3. File Organization

Organize documentation files systematically to facilitate easy access and retrieval.

11.3.4. Collaboration

Encourage collaboration among review team members to ensure all search-related activities are documented comprehensively.

11.4. Components of Reporting

11.4.1. PRISMA Flow Diagram

Include a PRISMA (Preferred Reporting Items for Systematic Reviews and Meta-

Analyses) flow diagram to illustrate the study selection process.

11.4.2. Search Strategy Description

Provide a detailed description of the search strategy, including search terms, databases, and any filters or limits applied.

11.4.3. Grey Literature Sources

List the sources of grey literature searched, along with the methods used to identify and include them.

11.4.4. Database Search Results

Report the total number of records retrieved from each database, and specify how duplicates were handled.

11.4.5. Study Selection Process

Describe the process of study selection, including the number of studies screened, assessed for eligibility, and included in the review.

11.4.6. Reasons for Exclusion

Provide a list of studies excluded during the full-text screening and the reasons for their exclusion.

11.4.7. Search Strategy Reproducibility

State whether the search strategy can be reproduced independently, and if possible, provide an appendix with the complete search strings.

11.5. Ensuring Quality in Reporting

11.5.1. Peer Review

Seek peer review of the search documentation and reporting to ensure accuracy and completeness.

11.5.2. Independent Validation

Encourage independent validation of the search strategy to confirm its reproducibility and comprehensiveness.

11.6. Challenges in Documentation and Reporting

11.6.1. Limited Space in Journals

Address space constraints in journal publications by providing additional documentation as supplementary material.

11.6.2. Language Barriers

Consider translating essential documentation and reporting into multiple languages to enhance accessibility.

11.6.3. Incomplete Reporting

Address any gaps or omissions in the documentation and reporting to ensure a comprehensive review process.

CHAPTER-12

SCREENING POTENTIALLY ELIGIBLE STUDIES: A COMPREHENSIVE REPORTING STANDARD

Once a systematic review's search process is complete, the next crucial step is the screening of potentially eligible studies. This chapter provides a comprehensive guide to the screening process, including developing a screening guide, conducting full-text screening, and adhering to reporting standards. Proper screening ensures that relevant studies are included, and the review remains transparent and reproducible. By following best practices and reporting standards, researchers can enhance the credibility and reliability of their systematic review.

12.1. Importance of Screening Potentially Eligible Studies

12.1.1. Ensuring Inclusion of Relevant Studies

The screening process is essential to identify and include studies that directly address the research question.

12.1.2. Minimizing Bias

A thorough screening process helps minimize selection bias and ensures the inclusion of all eligible studies.

12.1.3. Transparency and Reproducibility

Transparent reporting of the screening process allows other researchers to replicate and validate the study selection.

12.1.4. Compliance with Reporting Guidelines

Adhering to reporting guidelines, such as PRISMA (Preferred Reporting Items for Systematic Reviews and Meta-Analyses), ensures clarity and completeness in reporting the screening process.

12.2. Developing a Screening Guide

12.2.1. Formulating Inclusion and Exclusion Criteria

Define clear and explicit inclusion and exclusion criteria based on the research question and objectives.

12.2.2. Pilot Testing

Conduct a pilot screening phase to test the screening guide and refine the inclusion and exclusion criteria.

12.2.3. Training and Calibration

Train the review team on the screening guide and conduct calibration exercises to ensure consistency in screening decisions.

12.2.4. Handling Discrepancies

Establish a protocol for resolving disagreements among review team members during the screening process.

12.3. Title and Abstract Screening

12.3.1. Initial Screening

Screen titles and abstracts of retrieved records against the inclusion and exclusion criteria to identify potentially relevant studies.

12.3.2. Independent Screening

Ideally, conduct the initial screening independently by two or more review team members to enhance reliability.

12.3.3. Dual Review Process

Discuss and resolve discrepancies through a consensus-based approach or involve a third reviewer for resolution.

12.3.4. Recording Screening Decisions

Document the screening decisions for each study, including reasons for exclusion.

12.4. Full-Text Screening

12.4.1. Retrieving Full Texts

Obtain the full texts of potentially relevant studies identified during title and abstract screening.

12.4.2. Full-Text Review against Inclusion Criteria

Conduct a detailed assessment of each full text against the pre-defined inclusion and exclusion criteria.

12.4.3. Independent Full-Text Screening

Ideally, two or more review team members should independently conduct the full-text screening.

12.4.4. Data Extraction Planning

Prepare for data extraction by identifying key variables and information to be extracted from the included studies.

12.5. Reporting Standards for Screening Process

12.5.1. PRISMA Flow Diagram

Include a PRISMA flow diagram that illustrates the study selection process from identification to inclusion.

12.5.2. Screening Guide Description

Provide a clear and detailed description of the screening guide, including inclusion and exclusion criteria.

> **The screening of potentially eligible studies** is a pivotal step in conducting a systematic review. By developing a comprehensive screening guide, conducting thorough title and abstract screening, and employing independent full-text screening, researchers can ensure the inclusion of relevant studies and minimize bias. Transparent reporting and adherence to reporting standards, such as PRISMA guidelines, are essential to enhance the review's credibility and reproducibility. The screening process, when well-documented and meticulously conducted, plays a fundamental role in the systematic review's success, contributing to evidence-based decision-making and advancing knowledge in various fields of research.

12.5.3. Reasons for Exclusion

Report the reasons for excluding studies during both title and abstract screening and full-text screening.

12.5.4. Inter-Rater Agreement

Report the level of agreement among review team members during independent screening phases.

12.5.5. Full-Text Retrieval

Document the process of obtaining full-text articles, including communication with authors and publishers.

12.5.6. Discrepancy Resolution

Describe how discrepancies during screening were resolved, whether through consensus or involvement of a third reviewer.

12.6. Challenges and Limitations

12.6.1. Language Barriers

The inclusion of studies published in specific languages may pose challenges during the screening process.

12.6.2. Resource Constraints

Resource limitations, such as time and personnel, can impact the screening process's comprehensiveness and rigor.

12.6.3. Publication Bias

Publication bias may arise if certain studies are systematically excluded during the screening process.

PRISMA and the Generation of PRISMA Flow Diagram

The PRISMA (Preferred Reporting Items for Systematic Reviews and Meta-Analyses) statement is a widely recognized guideline that aims to enhance the transparency and reporting quality of systematic reviews. This chapter explores the significance of PRISMA in systematic reviews, its core components, and the step-by-step process of generating the PRISMA flow diagram. By adhering to PRISMA guidelines and utilizing the PRISMA flow diagram, researchers can present a comprehensive and transparent overview of the study selection process, thereby fostering credibility and reproducibility in their systematic reviews.

12.7.1. PRISMA: An Overview

12.7.1.1. Introduction to PRISMA

Introduce the PRISMA statement and its role in improving the reporting quality of systematic reviews.

12.7.1.2. The Evolution of PRISMA

Trace the development and history of PRISMA, highlighting its iterations and updates.

12.7.1.3. Benefits of PRISMA

Discuss the advantages of adhering to PRISMA guidelines, including improved transparency and reduced bias.

12.7.2. Core Components of PRISMA

12.7.2.1. Title and Abstract

Outline the key elements that should be included in the title and abstract of a systematic review.

12.7.2.2. Introduction

Explain the importance of providing a clear and concise introduction to the systematic review.

12.7.2.3. Methods

Detail the essential components of the methods section, including study design, search strategy, and study selection process.

12.7.2.4. Results

Discuss the presentation of results, including the characteristics of included studies and a summary of findings.

12.7.2.5. Discussion

Explore the significance of the discussion section, where researchers interpret the review's findings and discuss implications.

12.7.2.6. Funding and Conflicts of Interest

Explain the importance of disclosing funding sources and potential conflicts of interest in the systematic review.

12.7.3. The PRISMA Checklist

12.7.3.1. Structure of the PRISMA Checklist

Present a comprehensive overview of the PRISMA checklist, highlighting its key items and sections.

12.7.3.2. Importance of the PRISMA Checklist

Discuss the role of the PRISMA checklist in guiding researchers through the reporting process and promoting adherence to reporting standards.

12.7.3.3. Utilizing the PRISMA Checklist

Provide guidance on how to use the PRISMA checklist during the preparation and reporting of a systematic review.

12.7.4. The PRISMA Flow Diagram: Step-by-Step Guide

12.7.4.1. Purpose and Significance

Explain the purpose of the PRISMA flow diagram in visualizing the study selection process and summarizing included studies.

12.7.4.2. Components of the PRISMA Flow Diagram

Outline the key components of the PRISMA flow diagram, including the number of records identified, screened, assessed for eligibility, and included.

12.7.4.3. Step-by-Step Guide to Generating the PRISMA Flow Diagram

• Step 1: Describe the process of identifying and retrieving records through the search strategy.

• Step 2: Explain the title and abstract screening process, including the number of records screened and excluded.

• Step 3: Detail the full-text screening process, indicating the number of studies assessed for eligibility.

• Step 4: Present the reasons for excluding studies during full-text screening.

• Step 5: Summarize the number of studies included in the systematic review.

12.7.5. Reporting Standards for the PRISMA Flow Diagram

12.7.5.1. Transparent Reporting

Discuss the importance of transparently reporting the study selection process through the PRISMA flow diagram.

12.7.5.2. Completeness and Accuracy

Emphasize the significance of ensuring the completeness and accuracy of the PRISMA flow diagram.

12.7.5.3. Incorporating the PRISMA Flow Diagram in the Systematic Review Report

Provide guidance on how to include the PRISMA flow diagram in the systematic review report.

> **The PRISMA statement and the PRISMA flow diagram are** indispensable tools for improving the reporting quality and transparency of systematic reviews. By adhering to PRISMA guidelines and accurately generating the PRISMA flow diagram, researchers can present a clear and comprehensive summary of the study selection process, enhancing the credibility and reproducibility of their systematic reviews. The proper application of PRISMA contributes to evidence-based practice, informed decision-making, and the advancement of knowledge in various fields of research.

12.7.6. Interactive PRISMA Flow Diagram: PRISMA-P

12.7.6.1. Introduction to PRISMA-P

Introduce PRISMA-P, the extension of PRISMA for protocols of systematic reviews.

12.7.6.2. The Structure of PRISMA-P

Explain the components and items of PRISMA-P, guiding researchers in developing systematic review protocols.

12.7.6.3. Benefits of PRISMA-P

Discuss the advantages of utilizing PRISMA-P in protocol development to enhance the transparency and reproducibility of the systematic review process.

CHAPTER-13

GREY LITERATURE: SEARCHING, INCLUSION, AND SIGNIFICANCE

In the context of evidence synthesis, grey literature refers to research and information that is not published through traditional commercial publishing channels. It includes reports, theses, conference abstracts, working papers, government publications, and other unpublished or non-commercially published materials. This chapter explores the significance of grey literature in systematic reviews, the methods for searching and accessing it, and the considerations for its inclusion in systematic reviews. By incorporating grey literature, systematic reviews can achieve a more comprehensive and unbiased representation of the available evidence.

13.1. The Significance of Grey Literature in Systematic Reviews

13.1.1. Avoiding Publication Bias

Including grey literature helps reduce publication bias, as it encompasses research that may not be published in peer-reviewed journals.

13.1.2. Accessing Unpublished Data

Grey literature often contains valuable data and findings that may not be available through conventional sources.

13.1.3. Capturing Diverse Perspectives

Grey literature includes studies and reports from non-academic and non-traditional sources, providing a broader perspective on the research topic.

13.1.4. Timely Information

Grey literature may offer more recent and up-to-date information compared to peer-reviewed publications.

13.2. Sources of Grey Literature

13.2.1. Institutional Repositories

Many universities and research organizations maintain institutional repositories that host theses, dissertations, and other research outputs.

13.2.2. Government Agencies and Organizations

Government websites and publications often serve as valuable sources of grey literature, particularly in public policy and public health research.

13.2.3. Non-Governmental Organizations (NGOs)

NGOs and non-profit organizations frequently publish reports and research findings relevant to various fields.

13.2.4. Conference Proceedings

Conference abstracts and proceedings provide insights into preliminary research and emerging trends.

13.2.5. Industry Reports

Industry reports, technical documents, and white papers offer insights into corporate research and development.

13.2.6. Preprint Servers

Preprint servers host preliminary versions of research papers before peer review and publication.

13.3. Searching for Grey Literature

13.3.1. Comprehensive Search Strategies

Develop comprehensive search strategies to capture relevant grey literature, including specific search terms and filters.

13.3.2. Database Selection

Identify databases that specialize in grey literature, such as OpenGrey, Grey Literature Report, and BASE (Bielefeld Academic Search Engine).

13.3.3. Internet Search Engines

Utilize search engines like Google and Google Scholar to identify grey literature from websites and repositories.

13.3.4. Hand Searching

Conduct hand searches of relevant conference websites, institutional repositories, and organizational websites.

13.3.5. Expert Contacts

Reach out to experts in the field to inquire about unpublished research or grey literature sources.

13.4. Inclusion of Grey Literature in Systematic Reviews

13.4.1. Assessment of Relevance and Quality

Evaluate the relevance and quality of grey literature using similar criteria as applied to peer-reviewed publications.

13.4.2. Data Extraction and Synthesis

Extract relevant data from included grey literature and synthesize findings with those from peer-reviewed studies.

13.4.3. Transparent Reporting

Transparently report the process of searching, screening, and including grey literature in the systematic review.

13.5. Challenges and Limitations

Grey literature plays a pivotal role in systematic reviews, complementing the evidence from peer-reviewed publications and reducing publication bias. By adopting comprehensive search strategies, accessing diverse sources, and judiciously evaluating the relevance and quality of grey literature, researchers can enhance the robustness and validity of their systematic reviews. Transparent reporting and thoughtful inclusion of grey literature contribute to a more comprehensive and unbiased representation of the available evidence, enriching the review's findings and advancing evidence-based decision-making across various fields of research.

13.5.1. Access and Availability

Obtaining full-text access to some grey literature may be challenging due to copyright restrictions or limited availability.

13.5.2. Quality and Validity

Grey literature may vary in quality and validity, requiring careful assessment during inclusion.

13.5.3. Language Barriers

Grey literature may be published in languages other than English, posing potential language barriers.

13.5.4. Bias and Selective Reporting

The potential for bias and selective reporting exists in grey literature as well, necessitating cautious interpretation.

13.6. Advantages of Incorporating Grey Literature

13.6.1. Enhanced Scope

Including grey literature expands the scope of the systematic review, providing a more comprehensive overview of the research topic.

13.6.2. Real-World Relevance

Grey literature often reflects real-world experiences and practices, adding practical relevance to the review.

13.6.3. Addressing Research Gaps

Grey literature may help address gaps in the existing evidence base, providing insights into under-researched areas.

Quick source of Grey Literatures

Open Grey (www.opengrey.eu) for grey literature in Europe

Google Search to find any report/document

Social Science Research Network (http://ssrn.com/) is a research network for social sciences viz. economics, business management, psychology etc.

ProQuest and Shodhganga Reservoir of PhD theses

CHAPTER-14

DATA EXTRACTION AND CODING: TECHNIQUES FOR EFFECTIVE DATA MANAGEMENT AND DE-DUPLICATION

Data extraction and coding are fundamental steps in the systematic review process, ensuring that relevant information is extracted from selected studies and organized in a structured manner. This chapter delves into the techniques for data extraction and coding, discussing the types of data to extract and code from diverse sources.

Additionally, it explores the importance of data management and de-duplication to maintain the integrity and accuracy of the systematic review. By adhering to best practices and employing efficient data management strategies, researchers can confidently synthesize and analyze the extracted data to derive meaningful conclusions in their systematic reviews.

14.1. Data Extraction and Coding: An Overview

14.1.1. Purpose and Significance

Understand the purpose of data extraction and coding in a systematic review and its role in synthesizing evidence.

14.1.2. Data Management and Organization

Discuss the importance of structured data management and organization to facilitate data synthesis and analysis.

14.1.3. De-duplication

Introduce the concept of de-duplication and its role in eliminating duplicate records from the data set.

14.2. Types of Data to Extract and Code

14.2.1. Study Characteristics

Extract essential study characteristics, such as authors, publication year, study design, and sample size.

14.2.2. Intervention Details

Record details of the intervention, including the type, dosage, duration, and administration.

14.2.3. Outcome Measures

Extract outcome measures and metrics used in the studies to evaluate the intervention's effectiveness.

14.2.4. Effect Sizes and Confidence Intervals

Code effect sizes and confidence intervals to quantify the magnitude and precision of intervention effects.

14.2.5. Statistical Significance

Record information on statistical significance and p-values, aiding in the evaluation of study findings.

14.2.6. Adverse Events and Safety Data

Extract data on adverse events and safety outcomes associated with the intervention.

14.2.7. Follow-up Periods

Record the duration of follow-up periods to understand the long-term effects of the intervention.

> **Data extraction** and coding are critical components of a systematic review, enabling researchers to organize, synthesize, and analyze evidence effectively. By implementing efficient data management strategies and de-duplication methods, researchers can ensure the accuracy and reliability of the systematic review's findings. Transparent reporting of the data extraction and coding process enhances the review's credibility and reproducibility, contributing to evidence-based decision-making and advancing knowledge in various fields of research

14.3. Data Extraction and Coding Process

14.3.1. Formulating Data Extraction Forms

Develop data extraction forms based on the research question and predefined variables.

14.3.2. Calibration and Pilot Testing

Conduct calibration exercises and pilot testing to ensure consistency and accuracy among reviewers.

14.3.3. Independent Data Extraction

Assign two or more reviewers to independently extract data from selected studies to enhance reliability.

14.3.4. Resolving Discrepancies

Resolve discrepancies through discussion or consultation with a third reviewer when necessary.

14.3.5. Data Entry and Verification

Enter the extracted data into a secure database, ensuring accuracy and consistency during data entry.

14.4. Data Management and De-duplication

14.4.1. Database Management

Maintain a well-organized and structured database to facilitate data synthesis and analysis.

14.4.2. De-duplication Process

Implement de-duplication methods to eliminate duplicate records from the data set.

14.4.3. Manual De-duplication

Manually identify and remove duplicate records from the database.

14.4.4. Automated De-duplication

Utilize specialized software and tools to automate the de-duplication process.

14.4.5. Handling Multiple Publications

Address multiple publications of the same study to ensure data integrity.

14.5. Quality Assessment and Risk of Bias

14.5.1. Assessing Study Quality

Discuss methods for assessing the quality of included studies to evaluate the reliability of findings.

14.5.2. Evaluating Risk of Bias

Examine different dimensions of risk of bias and how it influences study validity.

14.5.3. Incorporating Quality Assessment in Data Analysis

Explore the impact of study quality on data synthesis and interpretation.

14.6. Data Synthesis and Meta-analysis

14.6.1. Approaches to Data Synthesis

Discuss different approaches to data synthesis, including narrative synthesis and meta-analysis.

14.6.2. Meta-analysis Techniques

Explain the process of conducting a meta-analysis and pooling effect sizes from multiple studies.

14.6.3. Forest Plots

Illustrate the results of meta-analysis using forest plots to visualize effect sizes and confidence intervals.

14.7. Reporting Standards for Data Extraction and Coding

14.7.1. PRISMA Data Extraction Checklist

Reference the PRISMA Data Extraction Checklist to ensure complete and transparent reporting.

14.7.2. Data Coding and Management Reporting

Outline the components to include in the systematic review report related to data extraction, coding, and management.

CHAPTER-15

CRITICAL APPRAISAL AND PUBLICATION BIAS: ENSURING RIGOR AND ADDRESSING BIASES

Systematic reviews are widely recognized as the gold standard for synthesizing evidence on specific research questions. However, their validity and reliability depend on the quality of the included studies and the potential for publication bias. This chapter delves into the critical appraisal process in systematic reviews and the importance of addressing publication bias to ensure rigor and minimize bias. By conducting thorough critical appraisal and implementing strategies to address publication bias, researchers can enhance the trustworthiness of their systematic reviews and improve evidence-based decision-making.

15.1. Critical Appraisal in Systematic Reviews

15.1.1. Defining Critical Appraisal

Explain the concept of critical appraisal in systematic reviews and its role in assessing study quality.

15.1.2. Importance of Critical Appraisal

Discuss the significance of critical appraisal in identifying strengths, weaknesses, and biases in included studies.

15.1.3. Types of Studies Included in Critical Appraisal

Outline the different types of studies (randomized controlled trials, cohort studies, case-control studies, etc.) included in systematic reviews and their respective critical appraisal considerations.

15.1.4. Tools for Critical Appraisal

Discuss commonly used critical appraisal tools, such as the Cochrane Risk of Bias tool, the Newcastle-Ottawa Scale, and the ROBINS-I tool.

15.1.5. Appraisal of Individual Study Characteristics

Detail the key study characteristics assessed during critical appraisal, including study design, sample size, blinding, allocation concealment, and others.

15.2. Conducting Critical Appraisal

15.2.1. Independent Assessment

Explain the importance of independent critical appraisal by two or more reviewers to enhance reliability.

15.2.2. Resolving Disagreements

Discuss strategies for resolving disagreements during critical appraisal, including involving a third reviewer if necessary.

15.2.3. Data Extraction and Risk of Bias Assessment

Explain the integration of critical appraisal findings into data extraction and risk of bias assessment.

15.2.4. Interpreting and Reporting Critical Appraisal Results

Provide guidance on how to interpret and transparently report the results of critical appraisal in systematic reviews.

Synthesizing Evidence: The Art of Systematic Review

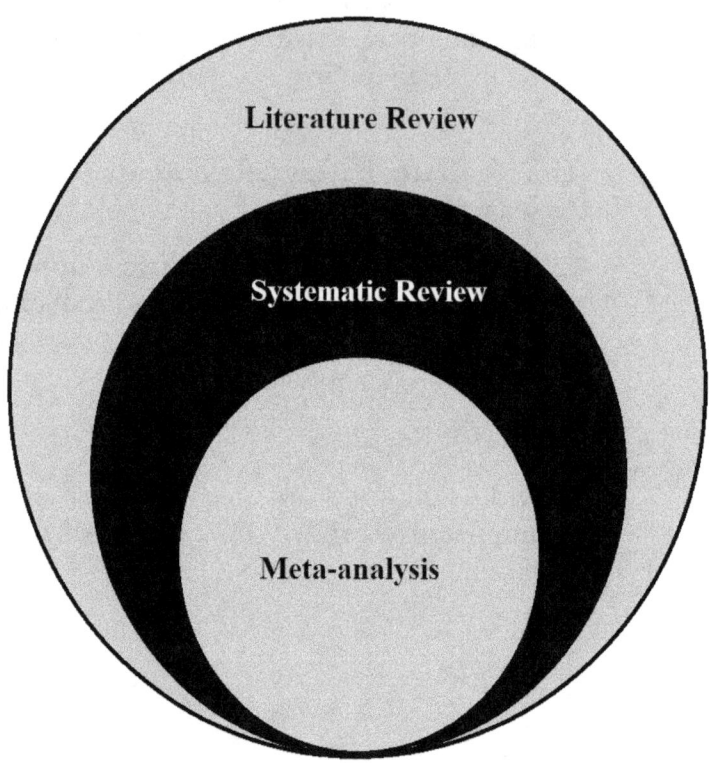

The relationships between the three pillars of evidence synthesis which are literature review, systematic review and meta-analysis

Cochrane Risk of Bias (RoB) tool

The Cochrane Risk of Bias (RoB) tool is a widely used and well-established tool designed to assess the risk of bias in randomized controlled trials (RCTs) included in systematic reviews. Developed by the Cochrane Collaboration, the RoB tool enables researchers to evaluate the methodological quality and internal validity of individual studies, which is crucial in determining the strength of evidence and reliability of the results in a systematic review.

Purpose of the Cochrane Risk of Bias Tool

The primary purpose of the Cochrane RoB tool is to assess the risk of bias in individual RCTs based on key domains that can potentially affect the validity of study results. By evaluating these domains, researchers can determine the likelihood of bias influencing the study outcomes and, consequently, the overall confidence in the findings of the systematic review.

Domains Assessed by the Cochrane Risk of Bias Tool

The Cochrane RoB tool evaluates the following key domains:

1. Random Sequence Generation: This domain assesses the method used to generate the random allocation sequence. Adequate randomization reduces the risk of selection bias.

2. Allocation Concealment: This domain examines the measures taken to conceal the treatment allocation sequence before assignment. Proper allocation concealment minimizes the risk of selection bias.

3. Blinding of Participants and Personnel: This domain assesses whether participants and personnel involved in the study were blinded to the intervention groups. Blinding helps reduce performance bias.

4. Blinding of Outcome Assessment: This domain evaluates whether outcome assessors were blinded to the intervention groups. Blinding of outcome assessment minimizes detection bias.

5. Incomplete Outcome Data: This domain assesses the extent to which outcome data are missing and the potential impact on the study results. Missing data can lead to attrition bias.

6. Selective Outcome Reporting: This domain examines whether the outcomes reported in the study align with the pre-specified outcomes. Selective outcome reporting can lead to reporting bias.

Scoring of the Cochrane Risk of Bias Tool

For each domain assessed by the Cochrane RoB tool, studies are categorized into one of three risk of bias levels:

• Low Risk of Bias: The study has taken appropriate measures to minimize bias in the specific domain, and the risk of bias is unlikely to significantly impact the study results.

• Unclear Risk of Bias: The study does not provide sufficient information to assess the risk of bias in the specific domain adequately.

• High Risk of Bias: The study has not adequately addressed bias in the specific domain, and the risk of bias is likely to impact the study results.

Using the Cochrane Risk of Bias Tool in Systematic Reviews

When conducting a systematic review, researchers use the Cochrane RoB tool to assess the risk of bias in each included RCT. The results of the risk of bias assessments are then considered during the data synthesis and interpretation stages of the review. Studies with a high risk of bias in critical domains may be given less weight in the overall analysis, and sensitivity analyses may be conducted to examine the impact of excluding studies with a high risk of bias.

The Cochrane Risk of Bias tool is a valuable instrument for evaluating the methodological quality and risk of bias in individual randomized controlled trials included in systematic reviews. By considering the risk of bias assessments, researchers can make informed decisions about the strength of evidence and confidently draw conclusions from the systematic review findings. The Cochrane RoB tool contributes to the credibility and validity of systematic reviews, enabling evidence-based decision-making in healthcare and policy domains.

Newcastle-Ottawa Scale (NOS)

The Newcastle-Ottawa Scale (NOS) is a widely used tool for assessing the quality of non-randomized studies, such as cohort studies and case-control studies, in systematic reviews and meta-analyses. It was developed by researchers from the University of Newcastle and the University of Ottawa and provides a systematic approach to evaluating the risk of bias and methodological quality of individual studies.

Purpose of the Newcastle-Ottawa Scale

The primary purpose of the Newcastle-Ottawa Scale is to assess the quality of non-randomized studies and to determine their risk of bias. By evaluating specific criteria related to participant selection, comparability of study groups, and outcome assessment, the NOS allows researchers to differentiate between high-quality and low-quality studies, which is crucial in synthesizing evidence in a systematic review.

Domains Assessed by the Newcastle-Ottawa Scale

The NOS evaluates the following key domains in non-randomized studies:

1. Selection of Study Groups (Cohort Studies) or Cases and Controls (Case-Control Studies): This domain assesses the representativeness of the study groups and the ascertainment of exposure or outcome status. It also considers the use of appropriate inclusion and exclusion criteria.

2. Comparability of Study Groups: For case-control studies, this domain evaluates whether cases and controls were adequately matched or adjusted for potential confounding factors. For cohort studies, it assesses whether the study groups were similar at the baseline.

3. Exposure/Outcome Assessment (Cohort Studies) or Exposure Assessment (Case-Control Studies): This domain examines the methods used to ascertain exposure or outcome status and the reliability of the assessment.

Scoring of the Newcastle-Ottawa Scale

The Newcastle-Ottawa Scale assigns stars or points to each study based on the criteria met in each domain. The maximum score for cohort studies is 9 points, while the maximum score for case-control studies is 10 points. Studies are categorized into one of three quality levels:

- High Quality (Low Risk of Bias): Studies with a high-quality score of 7 or more points (for cohort studies) or 8 or more points (for case-control studies) are considered to have a low risk of bias and are of high methodological quality.

• Moderate Quality: Studies with a moderate-quality score of 4 to 6 points (for cohort studies) or 5 to 7 points (for case-control studies) have a moderate risk of bias and are of moderate methodological quality.

• Low Quality (High Risk of Bias): Studies with a low-quality score of 3 or fewer points (for cohort studies) or 4 or fewer points (for case-control studies) are considered to have a high risk of bias and are of low methodological quality.

Using the Newcastle-Ottawa Scale in Systematic Reviews

During a systematic review, researchers apply the Newcastle-Ottawa Scale to assess the quality of each non-randomized study included in the review. The scores and quality levels of individual studies are considered when synthesizing evidence and

The Newcastle-Ottawa Scale is a valuable tool for evaluating the quality and risk of bias in non-randomized studies, such as cohort studies and case-control studies, included in systematic reviews. By considering the methodological quality of individual studies, researchers can assess the strength of evidence and confidently draw conclusions from the systematic review findings. The Newcastle-Ottawa Scale contributes to the credibility and validity of systematic reviews, allowing evidence-based decision-making in healthcare and policy domains.

drawing conclusions. High-quality studies are given more weight in the overall analysis, while low-quality studies may be excluded from sensitivity analyses to explore the robustness of the review's findings.

ROBINS-I tool

ROBINS-I is a tool designed to assess the risk of bias in non-randomized studies, particularly those evaluating the effects of interventions. Developed by the Cochrane Collaboration, the ROBINS-I tool provides a systematic and structured approach to evaluating the methodological quality and internal validity of non-randomized studies, such as cohort studies, before incorporating them into systematic reviews and meta-analyses.

Purpose of the ROBINS-I Tool

The primary purpose of the ROBINS-I tool is to assess the risk of bias in non-randomized studies that investigate interventions. It allows researchers to identify potential sources of bias in study design, conduct, and analysis, thus informing the confidence level in the results of these studies and their inclusion in systematic reviews.

Domains Assessed by the ROBINS-I Tool

The ROBINS-I tool assesses seven key domains, each representing a potential source of bias:

1. Bias Due to Confounding: This domain evaluates whether the study

adequately controlled for confounding variables through design or analysis.

2. Bias in Selection of Participants into the Study: This domain assesses the risk of bias related to the selection of participants and their exposure status.

3. Bias in Classification of Interventions: This domain examines whether there is a risk of misclassification of interventions or exposures.

4. Bias Due to Deviations from Intended Interventions: This domain evaluates whether there is a risk of bias due to deviations from the intended intervention.

5. Bias Due to Missing Data: This domain assesses the potential impact of missing data on the study results.

6. Bias in Measurement of Outcomes: This domain examines whether the outcome measurements are subject to bias.

7. Bias in Selection of Reported Results: This domain evaluates whether there is a risk of selective reporting of outcomes in the study.

Scoring of the ROBINS-I Tool

The ROBINS-I tool does not provide a specific scoring system like other bias assessment tools. Instead, it guides reviewers to assess the risk of bias in each domain and provide a narrative summary of the overall risk of bias for the study.

Using the ROBINS-I Tool in Systematic Reviews

> The ROBINS-I tool is a valuable instrument for evaluating the risk of bias in non-randomized studies, particularly those examining interventions. By systematically assessing key domains of bias, researchers can determine the methodological quality and internal validity of individual studies before incorporating them into systematic reviews. The ROBINS-I tool contributes to the credibility and validity of systematic reviews, enabling evidence-based decision-making in healthcare and policy domains, even when randomized controlled trials are not available or feasible.

During a systematic review, researchers use the ROBINS-I tool to assess the risk of bias in non-randomized studies that investigate interventions. The results of the risk of bias assessments are considered when synthesizing evidence and drawing conclusions. Studies with a low risk of bias are considered to have a higher internal validity and are given more weight in the overall analysis.

15.3. Publication Bias: Definition and Impact

15.3.1. Understanding Publication Bias

Define publication bias and explain how it can influence the findings of a systematic review.

15.3.2. Causes of Publication Bias

Discuss the various factors that contribute to publication bias, such as selective reporting and language bias.

15.3.3. Implications of Publication Bias

Explain the potential consequences of publication bias on the validity and generalizability of systematic review results.

15.4. Detecting and Assessing Publication Bias

15.4.1. Funnel Plots

Detail the use of funnel plots as a visual tool to detect publication bias.

Critical appraisal and addressing publication bias are essential components of rigorous systematic reviews. By thoroughly assessing study quality and implementing strategies to reduce bias, researchers can enhance the trustworthiness of their systematic review findings and contribute to evidence-based decision-making in various fields. Transparent reporting of critical appraisal results and measures to address publication bias strengthen the credibility and validity of systematic reviews, ensuring their meaningful impact on healthcare practice, policy development, and future research.

15.4.2. Statistical Tests

Discuss the use of statistical tests, such as Egger's test and Begg's test, to assess publication bias.

15.4.3. Sensitivity Analysis

Explain the role of sensitivity analysis in investigating the impact of publication bias on systematic review results.

15.5. Addressing Publication Bias

15.5.1. Comprehensive Literature Search

Discuss the importance of conducting a comprehensive literature search to minimize the risk of missing relevant studies.

15.5.2. Grey Literature and Unpublished Studies

Explore the inclusion of grey literature and unpublished studies in systematic reviews to reduce publication bias.

15.5.3. Trial Registration and Protocols

Discuss the significance of trial registration and protocols in addressing publication bias.

15.5.4. Contacting Authors

Explain the practice of contacting authors for additional data or unpublished studies.

15.6. Reporting Bias in Systematic Reviews

15.6.1. Selective Outcome Reporting

Discuss the potential for selective outcome reporting bias and strategies to address it.

15.6.2. Language Bias

Explain how language bias can lead to the exclusion of non-English studies and methods to mitigate it.

Funnel Plots: Visualizing Publication Bias and Small Study Effects

Funnel plots are graphical tools used in meta-analysis and systematic reviews to assess the potential presence of publication bias and small study effects. Asymmetry in funnel plots can indicate bias in the reporting and publication of studies, which may skew the overall estimate of treatment effects. This chapter explores the concept of funnel plots, their interpretation, and their role in evaluating the reliability and generalizability of meta-analysis results.

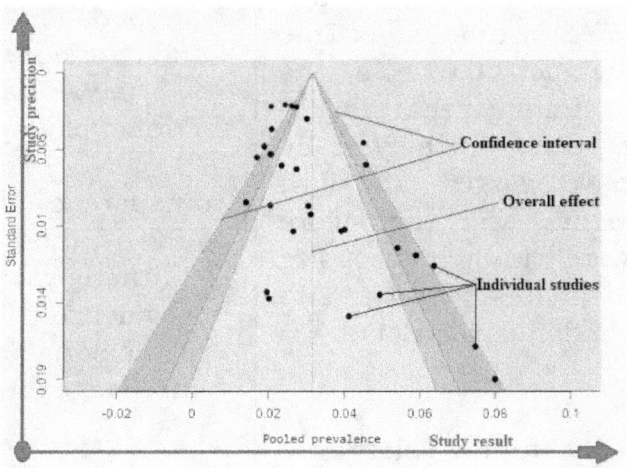

What are Funnel Plots?

Funnel plots are scatterplots that display the relationship between the effect size (e.g., odds ratio, standardized mean difference) and study precision (e.g., sample size, standard error) in meta-analyses.

Theoretical Background

Funnel plots are graphical tools commonly used in meta-analysis to assess publication bias, which occurs when the availability of research findings is influenced by the nature and direction of study results. In the absence of publication bias, funnel plots are expected to be symmetrical, with studies scattered evenly around the average effect size estimate. This symmetry reflects the assumption that smaller studies, which typically have larger standard errors, will vary more widely in their effect size estimates due to random error alone.

> Summarize the importance of funnel plots as visual tools for detecting publication bias and small study effects in meta-analyses. Emphasize the significance of careful interpretation and the consideration of additional methods to address biases and enhance the validity of meta-analysis results. Funnel plots play a crucial role in providing transparency and credibility to the synthesis of evidence, thereby contributing to evidence-based decision-making and advancing knowledge in various fields of research.

Why funnel plots are expected to be symmetrical in the absence of bias, could be understandable through several key concepts:

Sampling Variation: In meta-analysis, studies included in the analysis are typically a sample of all relevant studies that could have been conducted on a particular topic. Due to sampling variation, smaller studies are expected to vary more widely in their effect size estimates around the true underlying effect size.

Precision vs. Sample Size: Funnel plots typically plot the effect size estimate (e.g., standardized mean difference or odds ratio) against a measure of study precision (e.g., standard error or sample size). Smaller studies, with larger standard errors, are expected to have wider confidence intervals and thus will be more scattered around the average effect size estimate.

Absence of Bias: In the absence of bias, all studies, regardless of their precision, should have an equal chance of being published or included in the meta-analysis. This means that smaller studies with less precise estimates are equally likely to be included as larger studies with more precise estimates.

Publication Bias: When publication bias is present, it tends to affect smaller studies disproportionately. Negative or non-significant results from smaller studies may be less likely to be published or included in meta-analyses, leading to an asymmetric funnel plot where smaller studies with non-significant results are missing.

Detecting Publication Bias

Causes of Publication Bias

Publication bias refers to the tendency for research findings to be systematically published or disseminated based on the nature and direction of the results, leading to an incomplete or biased representation of the available evidence. Several factors contribute to publication bias, including:

Selective Publication of Studies with Positive Results: There is a well-documented tendency for studies with statistically significant or positive results to be more likely to be published than studies with non-significant or negative results. This phenomenon, known as the "file drawer problem," occurs because researchers, journal editors, and peer reviewers may perceive positive findings as more novel, interesting, or impactful, leading to a bias towards the publication of such studies.

Journal Bias: Some journals may prefer publishing studies with positive or statistically significant results, particularly those that are perceived to be more newsworthy or ground breaking. This bias can occur due to editorial policies, reviewer preferences, or commercial interests, leading to an overrepresentation of positive findings in the scientific literature.

Editorial Bias: Editorial decisions can be influenced by factors such as study design, sample size, and statistical significance, which may bias the publication process towards studies with positive results. Editors may prioritize studies that report novel or unexpected findings, leading to a bias against replication studies or studies with null results.

Funding Bias: Studies funded by industry or commercial interests may be more likely to report positive results or suppress negative findings in order to protect financial interests or maintain relationships with sponsors. This funding bias can contribute to the selective publication of studies with favourable outcomes and the underrepresentation of studies with unfavourable or neutral results.

Language Bias: Studies published in languages other than English may be less likely to be included in systematic reviews or meta-analyses, leading to a bias towards studies published in English-language journals. This language bias can result in an incomplete or skewed representation of the available evidence, particularly in fields where research is conducted globally.

Outcome Reporting Bias: Researchers may selectively report outcomes within a study based on their statistical significance or perceived importance, leading to a bias towards the publication of positive or significant outcomes. This outcome reporting bias can distort the interpretation of study findings and contribute to publication bias in the scientific literature.

Interpretation of Asymmetry

Asymmetry in funnel plots can be indicative of publication bias, suggesting that there may be a systematic relationship between the precision of study estimates and their effect sizes. Visual examination of funnel plot shapes can provide valuable insights into the presence and extent of publication bias. Here's how asymmetry in funnel plots can be interpreted in the context of publication bias:

Symmetrical Funnel Plot: In the absence of publication bias, funnel plots are expected to be symmetrical, with studies evenly distributed around the average effect size estimate. The scatter of points forms a funnel shape, with smaller studies scattered more widely due to the0ir larger standard errors. A symmetrical funnel plot indicates that there is no systematic relationship between study precision and effect size, suggesting that the available evidence is unbiased and represents a range of study sizes and findings.

Asymmetrical Funnel Plot - Egger's Test: Asymmetry in funnel plots, where there is a lack of smaller studies with non-significant or negative results, can be indicative of publication bias. One method for detecting asymmetry is Egger's test, which statistically assesses the relationship between study precision (e.g., standard error) and effect size. A significant p-value in Egger's test suggests that there is publication bias, with smaller studies tending to report more extreme or positive results.

Visual Examination of Funnel Plot Shape: Even without formal statistical tests, visual examination of funnel plot shapes can provide clues about the presence of publication bias. Asymmetry in the distribution of points, with fewer smaller studies on one side of the funnel plot, may suggest that smaller studies with non-significant or negative results are missing from the analysis. This can create a gap or skewness in the funnel plot, indicating the potential presence of publication bias.

Funnel Plot Regression: Another approach for detecting publication bias is funnel plot regression, which involves regressing effect size estimates against their standard errors. The slope of the regression line can provide an estimate of the degree of publication bias, with steeper slopes indicating more pronounced bias. If the regression line is significantly non-zero, it suggests that there is a systematic relationship between study precision and effect size, potentially indicating publication bias.

Statistical Tests for Publication Bias:

Statistical tests, such as Egger's test and Begg's test, are commonly used to assess funnel plot asymmetry and detect potential publication bias in meta-analyses. These tests evaluate whether there is a systematic relationship between the effect size estimates and their precision (e.g., standard error), which may indicate the presence of bias in the included studies. Here's a discussion of the use of these tests:

Egger's Test (Detailed are described later):

Method: Egger's test is a regression-based test that examines the association between the effect size estimates and their standard errors. It tests whether there is a significant relationship between the effect size and study precision, which would suggest asymmetry in the funnel plot.

Interpretation: A significant p-value in Egger's test indicates funnel plot asymmetry and provides evidence for the presence of publication bias. A non-significant p-value suggests no evidence of asymmetry and provides reassurance that publication bias is unlikely.

Considerations: Egger's test is sensitive to small-study effects and may have low power, particularly in meta-analyses with few included studies. It may also be influenced by heterogeneity and other sources of bias in the meta-analysis.

Begg and Majumdar's Test (Detailed are described later):

Method: Begg and Majumdar's test is a rank correlation test that evaluates the correlation between the effect size estimates and their ranks. It assesses whether there is a non-random relationship between study size and effect size, which may indicate publication bias.

Interpretation: A significant p-value in Begg and Majumdar's test suggests funnel plot asymmetry and provides evidence for publication bias. A non-significant p-value indicates no evidence of asymmetry and suggests that publication bias is unlikely.

Considerations: Begg and Majumdar's test is less sensitive to small-study effects compared to Egger's test and may be more appropriate when there are few studies or when the assumption of normality is violated. However, it may have lower power than Egger's test in detecting publication bias.

Other Methods:

Trim and Fill: Trim and Fill is a non-parametric method that imputes missing studies to assess and adjust for funnel plot asymmetry. It identifies potentially missing studies on one side of the funnel plot and estimates the effect of these studies on the overall meta-analysis results.

Regression-Based Approaches: Besides Egger's test, other regression-based methods, such as the "PET-PEESE" (Precision-Effect Estimate with Standard Error) method, can be used to assess and adjust for publication bias. These methods model the relationship between effect size estimates and their standard errors while accounting for heterogeneity and other sources of bias.

Small Study Effects

Small study effects refer to the phenomenon where smaller studies, typically characterized by smaller sample sizes, tend to produce effect size estimates that differ systematically from those of larger studies. This can manifest in various ways, such as smaller studies reporting larger effect sizes, more statistically significant results, or more extreme findings compared to larger studies. Small study effects can have a significant impact on the results of meta-analyses, and their presence may indicate underlying biases or methodological issues.

Causes of Small Study Effects

Small study effects can arise from various factors inherent to smaller studies and the publication process. Here are some common causes:

Publication Bias: Smaller studies with statistically significant or positive results are more likely to be published than those with non-significant or negative results.

This selective publication of studies can lead to an overrepresentation of positive findings in the literature, contributing to small study effects.

Selective Reporting Bias: Researchers and journals may be more inclined to report or publish studies with statistically significant or positive outcomes, while omitting non-significant or negative results. This selective reporting bias can skew the overall evidence base and influence meta-analyses.

Methodological Quality: Smaller studies may have methodological limitations or biases that affect the validity and reliability of their findings. These methodological flaws, such as inadequate sample sizes, lack of blinding, or incomplete outcome reporting, can lead to biased effect size estimates and contribute to small study effects.

Heterogeneity of Effects: Smaller studies may have more variability in their effect size estimates due to factors such as differences in study populations, interventions, or outcome measures. This heterogeneity can contribute to small study effects by influencing the overall estimate of the treatment effect in meta-analyses.

Publication Delays: Smaller studies may experience delays in publication or dissemination, particularly if they face challenges in securing funding, obtaining ethical approval, or completing data collection and analysis. These publication delays can affect the timing of study inclusion in meta-analyses and contribute to small study effects.

Language Bias: Studies published in languages other than English may be less likely to be included in systematic reviews and meta-analyses, leading to a potential bias towards studies published in English-language journals. This language bias can contribute to small study effects by limiting the representation of research findings from diverse linguistic and cultural contexts.

Data Dredging and Multiple Comparisons: In smaller studies, researchers may conduct multiple comparisons or exploratory analyses without adjusting for multiple testing, increasing the likelihood of chance findings and false-positive results. These data dredging practices can contribute to small study effects by inflating effect size estimates and statistical significance.

Egger's Test: Detecting Small Study Effects and Publication Bias in Meta-Analyses

Egger's test is a statistical method used in meta-analyses to assess the presence of small study effects and publication bias. It is named after the Swiss statistician **Matthias Egger**, who introduced the test in 1997. Here we explore the concept of Egger's test, its interpretation, and its role in evaluating the reliability and potential biases in meta-analysis results.

Introduction:

Egger's test is a statistical method used to assess publication bias in meta-analyses. Meta-analysis combines the results of multiple studies to provide a more precise estimate of the true effect size of an intervention or exposure. However, publication bias can distort the findings of meta-analyses if studies with significant results are more likely to be published than those with non-significant results. Egger's test provides a quantitative measure of funnel plot asymmetry, aiding researchers in identifying and correcting for publication bias.

Principles:

Egger's test assumes that in the absence of publication bias, the distribution of study effect sizes should resemble a symmetrical funnel shape in a scatterplot, with smaller studies scattered more widely at the bottom and larger studies clustered closer to the top. However, if there is publication bias, the funnel plot may exhibit asymmetry, with smaller studies reporting larger effect sizes than expected. Egger's test quantifies this asymmetry by regressing the standardized effect size (usually the log odds ratio or log risk ratio) against its standard error, with the intercept of the regression line indicating the degree of funnel plot asymmetry. A significant intercept suggests the presence of publication bias.

Applications:

Egger's test is widely used in systematic reviews and meta-analyses across various disciplines, including medicine, psychology, economics, and environmental science. It helps researchers assess the robustness of meta-analytic findings by detecting and quantifying the impact of publication bias. For example, in medical research, Egger's test is often used to evaluate the likelihood of selective publication of clinical trials based on their effect sizes and sample sizes.

Strengths:

One of the key strengths of Egger's test is its ability to provide a quantitative measure of publication bias, complementing the visual inspection of funnel plots. It offers a more objective assessment of asymmetry in the funnel plot and allows researchers to determine the statistical significance of publication bias. Additionally, Egger's test is relatively easy to implement and can be applied to meta-analyses with both continuous and binary outcome measures.

Limitations:

Despite its utility, Egger's test has several limitations that researchers should consider when interpreting its results. First, Egger's test requires a sufficient number of studies included in the meta-analysis to yield reliable estimates of publication bias. Small meta-analyses with fewer than ten studies may lack the statistical power to detect asymmetry in the funnel plot. Second, Egger's test assumes that publication bias is the only source of funnel plot asymmetry, which may not always be the case. Other factors, such as heterogeneity, selective outcome reporting, and methodological differences between studies, can also contribute to funnel plot asymmetry.

Interpretations:

The interpretation of Egger's test results depends on the statistical significance of the intercept in the regression analysis. A non-significant intercept suggests that there is no evidence of publication bias, and the observed asymmetry in the funnel plot is likely due to chance variation. Conversely, a significant intercept indicates the presence of publication bias, with smaller studies tending to report larger effect sizes than larger studies. In such cases, researchers should exercise caution when interpreting the findings of the meta-analysis and consider conducting sensitivity analyses to assess the robustness of the results.

Begg and Mazumdar's Test: Assessing Publication Bias in Meta-Analyses

Begg and Majumdar's test is a statistical method commonly used in meta-analyses to evaluate the presence of publication bias. It was introduced by Colin B Begg and Madhuchhanda Mazumdar in 1994 and is designed to complement other tools, such as Egger's test and visual examination of funnel plots, in assessing the reliability and potential biases in meta-analysis results. Here we explore the concept of Begg and Majumdar's test, its interpretation, and its role in detecting publication bias in systematic reviews.

Introduction:

Begg and Majumdar's test is a statistical method used in meta-analysis to assess publication bias, a phenomenon where studies with significant results are more likely to be published than those with non-significant results. Publication bias can distort the findings of meta-analyses, leading to biased effect size estimates and potentially erroneous conclusions. Begg and Majumdar's test offers a quantitative approach to detecting publication bias by examining the correlation between study effect sizes and their variances.

Principles:

Begg and Majumdar's test assumes that in the absence of publication bias, there should be no correlation between study effect sizes and their variances. In other words, studies with larger variances (i.e., greater uncertainty) should not systematically report larger or smaller effect sizes than studies with smaller variances. Begg and Majumdar's test evaluates this assumption by calculating the rank correlation coefficient (Kendall's tau) between the ranks of effect sizes and the ranks of their variances across studies. A significant correlation suggests the presence of publication bias.

Applications:

Begg and Majumdar's test is commonly used in systematic reviews and meta-analyses across various fields, including medicine, psychology, social sciences, and environmental science. It serves as a valuable tool for assessing the robustness of meta-analytic findings and identifying potential biases that may affect the validity of conclusions drawn from the synthesis of research evidence. Researchers often employ Begg and Majumdar's test in conjunction with other methods, such as

Egger's test and funnel plot inspection, to comprehensively evaluate publication bias.

Strengths:

One of the key strengths of Begg and Majumdar's test is its ability to detect publication bias without relying on assumptions about the distribution of effect sizes or the shape of the funnel plot. Unlike Egger's test, which assesses funnel plot asymmetry using regression analysis, Begg and Majumdar's test directly examines the relationship between effect sizes and their variances, making it less susceptible to confounding factors. Additionally, Begg and Majumdar's test is relatively straightforward to implement and can be applied to meta-analyses with both continuous and binary outcome measures.

Limitations:

Despite its utility, Begg and Majumdar's test has certain limitations that researchers should consider when interpreting its results. Like Egger's test, Begg and Majumdar's test requires a sufficient number of studies included in the meta-analysis to yield reliable estimates of publication bias. Small meta-analyses with fewer than ten studies may lack the statistical power to detect publication bias using Begg and Majumdar's test. Furthermore, Begg and Majumdar's test may be less sensitive to subtle forms of publication bias compared to other methods, such as Egger's test, which directly assess funnel plot asymmetry.

Interpretations:

The interpretation of Begg and Majumdar's test results hinges on the statistical significance of the rank correlation coefficient (Kendall's tau). A non-significant correlation suggests that there is no evidence of publication bias, and the observed relationship between effect sizes and their variances is likely due to chance variation. Conversely, a significant correlation indicates the presence of publication bias, with studies reporting larger effect sizes having larger variances, or vice versa. In such cases, researchers should exercise caution when interpreting the findings of the meta-analysis and consider conducting sensitivity analyses to assess the robustness of the results.

CHAPTER-16

RANDOMIZED CONTROLLED TRIALS (RCTS): AN ADVANCE IN EVIDENCE-BASED PRACTICE

Randomized Controlled Trials (RCTs) represent a cornerstone in the field of scientific research, particularly in medicine, healthcare, and social sciences. These meticulously designed experiments aim to evaluate the efficacy and safety of interventions or treatments by comparing outcomes between groups of participants who are randomly assigned to receive different interventions. Concurrently, systematic reviews, which involve a comprehensive and systematic synthesis of existing evidence, including data from RCTs, play a pivotal role in summarizing and evaluating the cumulative body of knowledge on a particular topic. This essay explores the fundamental principles of RCTs, their significance in evidence-based practice, and the impact of systematic reviews in synthesizing RCT findings to inform decision-making in healthcare and beyond.

16.1. Understanding Randomized Controlled Trials

Randomized Controlled Trials are experimental studies characterized by several key features:

Randomization: Participants are randomly assigned to either an experimental group or a control group. Random allocation helps ensure that each group is comparable in terms of baseline characteristics, minimizing selection bias and confounding variables.

Control Group: The control group serves as a reference point and may receive either no intervention (placebo) or standard treatment. Comparing outcomes between the experimental and control groups allows researchers to evaluate the efficacy of the intervention being studied.

Blinding: Blinding, or masking, is often employed to minimize bias. In single-blind studies, participants are unaware of their treatment assignment, while in double-blind studies, both participants and researchers are unaware. Blinding helps prevent subjective influences on outcome assessments.

Outcome Measures: RCTs typically include predefined outcome measures to assess the effects of the intervention. These outcomes may include clinical endpoints (e.g., improvement in symptoms, disease progression) or surrogate endpoints (e.g., biomarkers).

Sample Size: Adequate sample size is crucial to ensure the statistical power of the study. Power calculations are performed to determine the number of participants needed to detect clinically meaningful differences between groups with statistical significance.

Ethical Considerations: RCTs must adhere to ethical principles, including informed consent, protection of participant confidentiality, and minimizing risks to participants. Institutional review boards oversee the ethical conduct of RCTs to ensure participant safety and welfare.

16.2. Significance of Randomized Controlled Trials

Randomized Controlled Trials hold significant importance in advancing evidence-based practice for several reasons:

Causality: RCTs provide strong evidence of causality by demonstrating the efficacy of interventions through rigorous experimental design. Randomization helps minimize bias, allowing researchers to establish a causal relationship between the intervention and outcome.

Clinical Decision Making: Findings from well-conducted RCTs inform clinical decision-making by providing healthcare practitioners with reliable evidence on the effectiveness and safety of interventions. Clinicians can use RCT results to make informed treatment decisions tailored to individual patient needs.

Policy Development: RCTs contribute to the development of healthcare policies and guidelines by providing robust evidence on the efficacy and cost-effectiveness of interventions. Policy-makers rely on RCT findings to formulate evidence-based strategies for improving population health outcomes.

Innovation and Research Progression: RCTs drive innovation and research progression by evaluating novel interventions and treatments. By identifying effective interventions and refining existing practices, RCTs contribute to the advancement of medical science and patient care.

16.3. Impact of Systematic Reviews on Evidence-Based Practice

Systematic reviews have a profound impact on evidence-based practice in several ways:

Summarizing and Synthesizing Evidence: Systematic reviews provide clinicians, researchers, and policymakers with a consolidated summary of the available evidence on a particular topic. By synthesizing data from multiple studies, systematic reviews offer a comprehensive overview of the efficacy, safety, and limitations of interventions.

Informing Clinical Guidelines and Practice Recommendations: Findings from systematic reviews inform the development of clinical guidelines and practice recommendations. Professional organizations and healthcare institutions rely on systematic reviews to establish evidence-based guidelines for optimal patient care.

Identifying Evidence Gaps and Priorities for Research: Systematic reviews identify gaps in the existing evidence base and highlight areas where further research is needed. By identifying research priorities, systematic reviews guide the allocation of resources and funding to address critical knowledge gaps.

Facilitating Shared Decision Making: Systematic reviews empower patients and healthcare providers to engage in shared decision-making by providing access to evidence-based information on treatment options, benefits, and risks. Informed decision-making enhances patient satisfaction and improves health outcomes.

16.4. Enhancing Policy Development and Resource Allocation: Policymakers and healthcare administrators use systematic reviews to inform policy development and resource allocation decisions. Evidence-based policies grounded in systematic reviews promote cost-effective interventions and improve healthcare delivery.

Advancing Research Methodology: Systematic reviews contribute to the advancement of research methodology by establishing standards for evidence synthesis, meta-analysis, and quality assessment. Methodological advancements enhance the rigor and transparency of systematic reviews, further strengthening their impact on evidence-based practice.

16.5. Differences Between a Randomized-Controlled Trial vs Systematic Review

Policymakers and system implementers bear the responsibility of making pivotal decisions that will shape current and future strategies. To ensure these decisions are both effective and unbiased, they must be grounded in solid, incontrovertible, and relevant evidence. Consequently, a comprehensive review of numerous studies is imperative to gather sufficient data to inform these crucial decisions. Failure to review all pertinent eligible studies may result in inconsistent or incomplete results. This underscores the importance of various types of reviews, including systematic reviews and integrated reviews, among others.

Through systematic reviews, stakeholders can assess available literature and compile credible data to serve as evidence for decision-making. However, it is essential to comprehend the strengths and limitations of each review type to select the most suitable one based on specific objectives. Fortunately, abundant online resources elucidate the distinctions between integrated reviews versus systematic reviews, rapid reviews versus systematic reviews, and other comparative analyses.

A systematic review is guided by a clearly defined research question and employs a systematic and reproducible methodology to identify, select, and analytically evaluate all relevant research. It meticulously gathers and analyzes eligible studies from reputable research sources to bolster the evidence base. A crucial aspect of the systematic review process is ensuring that it addresses a specific research question. The research question, along with the study objectives and topic, delineates the scope of the review, safeguarding against deviation from its intended purpose.

To embark on a systematic review, it is imperative to develop and register a review protocol, outlining the rationale and eligibility criteria. This protocol serves as a blueprint for the review process, pre-empting potential issues or discrepancies in the results. Additionally, it enables transparency and reproducibility by providing insight into the review methodology for other reviewers. Consequently, the systematic review protocol enhances the review's credibility and facilitates quality assessment.

A Randomized Controlled Trial (RCT) is a scientifically designed trial aimed at controlling variables that are not directly under experimental control. Typically used in medical research, a prime example of an RCT is a clinical trial assessing the effects of pharmacological treatments, surgical procedures, medical devices, diagnostic techniques, or other medical interventions.

In an RCT, participants (or subjects) are randomly assigned to either experimental groups (EG) or control groups (CG), ensuring impartiality in group assignments. The hallmark of RCTs lies in the complete randomization of participant allocation to EG or CG. This allocation may be single-blinded, double-blinded, or unblinded. In a double-blind RCT, neither participants nor assessment professionals know the group assignments, minimizing potential biases and ensuring objective outcomes. The experimental group receives the intervention or treatment, while the control group may receive a placebo, an alternative treatment, or no treatment at all. Double-blinding ensures that no one can influence the results, enhancing the study's integrity.

RCTs offer several advantages and disadvantages. On the one hand, RCTs mitigate population biases, providing impartial evidence for informed decision-making. Unlike observational studies, the blinding in RCTs prevents subjectivity in outcome assessments, enhancing the validity of results. Furthermore, RCTs allow for the use of established statistical tools, facilitating result analysis and interpretation.

However, RCTs can be resource-intensive due to the need for a large participant pool and extended follow-up periods for comprehensive analysis. Nonetheless, cost-effective measures, such as employing simple outcome measures, can help manage RCT expenses.

Similar to how an RCT requires predefined criteria for participant inclusion, conducting a systematic review necessitates clear standards for research inclusion. While RCTs are esteemed for their robust design, they may not address all clinical inquiries, prompting the inclusion of observational research in study designs as needed. Thus, integrating various study designs, including RCTs and observational studies, ensures a comprehensive approach to evidence generation and decision-making.

16.6. Advantages of RCTs:

Minimization of Bias: Random allocation of participants to treatment and control groups helps minimize selection bias and confounding variables. This ensures that the groups are comparable at baseline, enhancing the internal validity of the study.

Causality Establishment: RCTs allow researchers to establish causal relationships between interventions and outcomes. By controlling for other variables, RCTs provide strong evidence of cause and effect, supporting evidence-based decision-making.

High Internal Validity: RCTs are designed to maximize internal validity, reducing the likelihood of alternative explanations for observed effects. Blinding techniques further enhance internal validity by minimizing bias in outcome assessments.

Objective Outcome Assessment: Blinding in RCTs ensures that outcome assessments are objective and unbiased, reducing the risk of subjective interpretation of results.

Statistical Rigor: RCTs use robust statistical methods to analyze data, providing quantitative measures of intervention effects. This enhances the reliability and reproducibility of study findings.

Generalizability: Well-designed RCTs with diverse participant samples can yield findings that are generalizable to broader populations, informing clinical practice and policy decisions.

16.7. Disadvantages of RCTs:

Cost and Resource Intensiveness: RCTs can be expensive and time-consuming due to the need for large sample sizes, long follow-up periods, and comprehensive data collection and analysis.

Ethical Considerations: Some interventions tested in RCTs may pose ethical dilemmas, particularly if there is uncertainty about potential harms or benefits to participants.

Limited External Validity: Strict inclusion and exclusion criteria in RCTs may limit the generalizability of findings to real-world settings or specific patient populations.

Practical Challenges: Recruitment and retention of participants, adherence to treatment protocols, and compliance with study procedures can pose practical challenges in conducting RCTs.

Infeasibility for Certain Research Questions: RCTs may not be feasible or ethical for certain research questions, particularly those related to rare diseases, long-term outcomes, or complex interventions.

Publication Bias: Positive results from RCTs are more likely to be published than negative or inconclusive results, leading to publication bias and potentially biased evidence synthesis.

In summary, while Randomized Controlled Trials offer robust evidence for evaluating interventions, researchers must carefully weigh the advantages and disadvantages to determine the appropriateness of an RCT for addressing specific research questions and informing evidence-based practice.

Randomized Controlled Trials and Systematic Reviews are essential components of evidence-based practice, playing complementary roles in generating, synthesizing, and evaluating research evidence. RCTs provide rigorous experimental evidence of intervention efficacy, while systematic reviews offer comprehensive syntheses of the collective evidence from multiple studies. Together, RCTs and systematic reviews drive advancements in healthcare, inform clinical decision-making, and shape policy and practice guidelines. By adhering to rigorous methodologies and upholding principles of transparency and integrity, RCTs and systematic reviews contribute to the continuous improvement of patient care and research outcomes.

CHAPTER-17

DISSEMINATING SYSTEMATIC REVIEW FINDINGS: PERFORMING EVIDENCE SYNTHESIS

Systematic reviews are essential in evidence-based practice, providing comprehensive and unbiased summaries of existing research on specific topics. However, the value of a systematic review lies not only in its creation but also in its dissemination. This chapter explores the significance of disseminating systematic review findings and performing evidence synthesis to effectively communicate research outcomes to various stakeholders. By employing diverse dissemination strategies and synthesizing evidence, researchers can enhance the impact of their systematic reviews and promote evidence-based decision-making in clinical practice, policy development, and future research.

17.1. Introduction:

Systematic reviews play a crucial role in synthesizing existing evidence on a particular topic, providing valuable insights for decision-makers, practitioners, and researchers. However, the dissemination of systematic review findings is essential to ensure that the synthesized evidence reaches its intended audience and has a meaningful impact on policy, practice, and further research. Performing evidence synthesis involves a series of steps aimed at effectively communicating the results of a systematic review to diverse stakeholders.

17.2. Steps in Performing Evidence Synthesis:

17.2.1. Summarizing Key Findings:

The first step in evidence synthesis is to summarize the key findings of the systematic review concisely and clearly. This involves identifying the main outcomes, effect sizes, and implications of the synthesized evidence. Summaries should be tailored to different audiences, such as policymakers, healthcare providers, and the general public, to ensure relevance and accessibility.

17.2.2. Creating Visual Aids:

Visual aids, such as tables, figures, and infographics, can help convey complex information in a visually appealing and understandable format. These visual aids can be used to illustrate key findings, trends, and relationships identified in the systematic review. Care should be taken to design visual aids that are clear, intuitive, and informative.

17.2.3. Developing Executive Summaries:

Executive summaries provide a high-level overview of the systematic review findings, making them particularly useful for busy decision-makers and stakeholders. Executive summaries should highlight the most important findings, implications, and recommendations of the systematic review in a concise and accessible format.

17.2.4. Preparing Policy Briefs and Practice Guidelines:

Policy briefs and practice guidelines translate systematic review findings into actionable recommendations for policymakers, healthcare providers, and other stakeholders. These documents outline evidence-based practices, interventions, and policies informed by the synthesized evidence, helping to guide decision-making and improve outcomes.

17.2.5. Disseminating Through Peer-Reviewed Publications:

Publishing the systematic review findings in peer-reviewed journals ensures that the research undergoes rigorous peer review and reaches a wide audience of researchers and practitioners. Authors should select journals with relevant scope and readership and follow the submission guidelines to maximize the impact and visibility of their work.

17.2.6. Presenting at Conferences and Workshops:

Presenting the systematic review findings at conferences, workshops, and professional meetings provides an opportunity to engage with peers, stakeholders, and policymakers directly. Presenters should use clear and engaging visuals, narrative storytelling, and interactive formats to communicate their findings effectively and facilitate discussion.

17.2.7. Engaging with Stakeholders:

Engaging with stakeholders throughout the systematic review process ensures that the research addresses their needs, priorities, and concerns. Stakeholder engagement can take various forms, including advisory panels, focus groups, interviews, and surveys, and should be tailored to the preferences and interests of different stakeholder groups.

17.2.8. Utilizing Social Media and Online Platforms:

Leveraging social media platforms, blogs, and websites can enhance the visibility and reach of systematic review findings. Authors can share summaries, infographics, and key messages from the systematic review on platforms such as Twitter, LinkedIn, ResearchGate, and institutional websites to engage with a broader audience and facilitate knowledge dissemination.

CHAPTER-18

REFERENCE MANAGEMENT

Systematic reviews are comprehensive and meticulous methods of synthesizing research evidence, crucial for informing evidence-based practices across various fields. One fundamental component of conducting a systematic review is effective reference management. This involves organizing, storing, and citing numerous references in a way that ensures accuracy, efficiency, and reproducibility. Here, we delve into the importance of reference management, best practices, tools, and strategies for maintaining high standards in systematic reviews.

18.1. The Importance of Reference Management

Effective reference management is essential for several reasons:

18.1.1. Accuracy and Integrity:
Accurate reference management ensures that all sources are correctly cited, preventing issues related to plagiarism and maintaining the integrity of the research.

18.1.2. Efficiency:
Systematic reviews often involve hundreds, if not thousands, of references. Efficient management saves time and reduces the risk of errors.

18.1.3. Reproducibility:
Good reference management practices facilitate the reproducibility of systematic reviews, allowing other researchers to verify and build upon the work.

18.1.4. Compliance:
Properly managing references helps comply with the guidelines of publishers and institutions, which often have strict requirements for citation formats and reference lists.

18.2. Best Practices in Reference Management

To achieve effective reference management in systematic reviews, researchers should adhere to several best practices:

18.2.1. Early Planning:
Plan your reference management strategy early in the review process. Decide on the tools and methods you will use to collect, organize, and cite references.

18.2.2. Consistent Use of Citation Styles:
Adhere to a consistent citation style as required by your target publication or

institution. Popular styles include APA, MLA, Chicago, and Harvard.

18.2.3. Regular Updates: Regularly update your reference list as you add new sources. This helps in keeping track of all the references and avoids last-minute rushes.

18.2.4. Comprehensive Records: Maintain comprehensive records of all references, including full bibliographic details and access information.

18.2.5. Deduplication: Systematically check for and remove duplicate references to maintain a clean and organized reference list.

18.3. Tools for Reference Management

Several tools can aid in reference management for systematic reviews. These tools range from simple databases to sophisticated software that can handle large volumes of data and complex citation needs. Here are some of the most widely used tools:

18.3.1.: EndNote is a popular reference management software that allows users to organize references, create bibliographies, and integrate with word processors for easy citation insertion. It supports various citation styles and can handle large reference libraries.

18.3.2. Mendeley: Mendeley combines reference management with social networking for researchers. It allows for easy organization of references, collaboration with other researchers, and discovery of new research.

18.3.3. Zotero: Zotero is a free, open-source reference manager that is user-friendly and integrates well with web browsers and word processors. It supports a wide range of citation styles and offers robust features for organizing and sharing references.

18.3.4. RefWorks: RefWorks is a web-based reference management tool that facilitates the collection, organization, and citation of research materials. It is particularly useful for collaborative projects due to its sharing capabilities.

18.3.5. COVIDENCE: Covidence is specifically designed for managing systematic reviews. It helps streamline the process of screening, data extraction, and reference management, making it an invaluable tool for systematic review authors.

18.4. Strategies for Effective Reference Management

Beyond selecting the right tools, employing effective strategies is crucial for managing references in systematic reviews. These strategies ensure that the reference management process is smooth and that all necessary details are meticulously recorded.

18.4.1. Develop a Workflow: Establish a clear workflow for reference management. This includes defining stages such as literature search, reference importation, screening, data extraction, and citation.

18.4.2. Use Tags and Folders: Organize references using tags and folders. This helps in quickly locating and categorizing references based on themes, stages of review, or relevance.

Automate Where Possible: Utilize automation features in reference management tools to import references directly from databases, generate citations, and check for duplicates.

18.4.3. Collaborate with Co-Authors: For multi-author systematic reviews, ensure that all team members are aligned on the reference management process. Use shared libraries and collaborative tools to maintain consistency.

18.4.4. Regular Backups: Regularly back up your reference library to prevent data loss. Most reference management tools offer cloud storage options for easy backup and retrieval.

18.4.5. Detailed Notes: Keep detailed notes on each reference, including summaries, relevance to the review, and any methodological details. This aids in the data extraction and synthesis phases.

18.5. Implementing Reference Management in a Systematic Review

Implementing effective reference management in a systematic review involves several key steps, which are as follows:

Initial Setup:

> Choose a reference management tool that suits your needs.

> Set up your reference library, including folders and tags for organization.

Conducting Literature Search:

> Perform a comprehensive literature search using multiple databases.

> Import references into your reference management tool directly from databases using compatible plugins or export features.

Screening and Selection:

> Screen references for relevance according to predefined inclusion and exclusion criteria.

> Use your reference management tool to tag and organize references that meet the criteria.

Data Extraction and Synthesis:

> Extract relevant data from selected references.

Use notes and tags to keep track of extracted information and its sources.

Writing and Citing:

As you write your systematic review, use your reference management tool to insert citations.

Ensure that all citations are accurate and formatted according to the required citation style.

Finalizing the Review:

Before final submission, double-check your reference list for completeness and accuracy.

Use the reference management tool's built-in features to generate the bibliography and ensure consistency.

18.6. Challenges and Solutions

Despite the availability of advanced tools and strategies, reference management in systematic reviews can pose several challenges. Here are common challenges and solutions:

18.6.1. Handling Large Volumes of References:

Solution: Use powerful reference management tools capable of handling large datasets, and employ filters and tags to manage them efficiently.

18.6.2. Ensuring Data Accuracy:

Solution: Regularly cross-check references and use built-in validation features in reference management tools.

18.6.3. Maintaining Consistency Across Multiple Authors:

Solution: Establish clear guidelines and workflows for all authors, and use collaborative features of reference management tools to synchronize efforts.

18.6.4. Dealing with Duplicate References:

Solution: Use automated deduplication features and regularly review your reference library for duplicates.

CHAPTER-19

TECHNOLOGICAL ADVANCEMENT AND APPLICATIONS OF MACHINE LEARNING

The field of systematic reviews has witnessed significant advancements in recent years, thanks to the rapid development of technology and machine learning techniques. These innovations have revolutionized the way systematic reviews are conducted, making the process more efficient, accurate, and accessible.

This chapter explores the various technological advances and machine learning applications in systematic reviews. From literature screening to data extraction and evidence synthesis, machine learning plays a vital role in automating tasks and improving the quality and speed of systematic reviews.

19.1. Automated software to conduct systematic reviews

Several software tools are available to assist researchers in conducting systematic reviews. Here are some popular options:

19.1.1. COVIDENCE: Covidence is a web-based platform specifically designed for managing systematic reviews and streamlining the review process. It facilitates collaboration among reviewers, automates screening and data extraction, and provides tools for risk of bias assessment and data synthesis.

19.1.2. DistillerSR: DistillerSR is another web-based systematic review management tool that offers features such as automated deduplication, customizable screening forms, data extraction templates, and risk of bias assessment tools. It also provides advanced reporting and visualization options.

19.1.3. EPPI-Reviewer: EPPI-Reviewer is a software tool developed by the Evidence for Policy and Practice Information and Coordinating Centre (EPPI-Centre) for conducting systematic reviews and meta-analyses. It offers features for screening, data extraction, quality assessment, and synthesis of evidence.

19.1.4. RevMan (Review Manager): RevMan is a desktop-based software developed by the Cochrane Collaboration for preparing and maintaining Cochrane Reviews. It provides templates for writing review protocols and full reviews, as well as tools for conducting meta-analyses and presenting results.

19.1.5. MetaXL: MetaXL is an add-in for Microsoft Excel that facilitates meta-analysis of summary data from systematic reviews and meta-analyses. It provides tools for conducting fixed-effect and random-effects meta-analyses, assessing heterogeneity, and generating forest plots.

19.1.6. Rayyan: Rayyan is a web-based application for screening and managing systematic review literature. It offers features for collaboration, automated deduplication, screening, and data extraction. It also provides integration with citation management tools such as EndNote and Zotero.

19.1.7. JBI SUMARI: JBI SUMARI (System for the Unified Management, Assessment, and Review of Information) is a software tool developed by the Joanna Briggs Institute for conducting systematic reviews of evidence generated using a range of review methodologies.

19.1.8. Mendeley: Mendeley is a reference management tool that can also be used for organizing and annotating systematic review literature. It offers features for collaboration, document tagging, and full-text search, making it useful for managing large sets of references.

Besides, these tools there are different search engine or computational tool or database that utilizes artificial intelligence (AI) to provide comprehensive and contextually relevant search results.

19.2. AI supported tools

19.2.1. SEMANTIC SCHOLAR

Semantic Scholar is an academic search engine and database that utilizes artificial intelligence (AI) techniques to provide comprehensive and contextually relevant search results in the field of scientific research. Developed by the Allen Institute for Artificial Intelligence (AI2), Semantic Scholar aims to streamline the process of accessing scholarly literature and extracting valuable insights from vast amounts of scientific data. Key features of Semantic Scholar are as follows:

Semantic Search: Semantic Scholar employs advanced natural language processing (NLP) algorithms to understand the meaning and context of academic papers. This enables it to deliver highly relevant search results that match the user's query, even if the exact keywords are not present in the document.

AI-Powered Recommendations: The platform utilizes machine learning algorithms to analyze user behavior and preferences, providing personalized recommendations for relevant research papers, authors, and topics.

Citation Analysis: Semantic Scholar incorporates citation analysis to identify influential papers, authors, and research trends within specific fields or disciplines. This helps researchers discover seminal works and track the impact of their own research.

Contextual Insights: Semantic Scholar provides additional context and metadata for research papers, including citation counts, publication venues, author affiliations, and related works. This enriches the reading experience and facilitates deeper exploration of scientific literature.

Open Access Integration: The platform integrates with various open access repositories and publishers to ensure access to a wide range of scholarly content. Users can access full-text articles directly through Semantic Scholar or via external links to publisher websites.

19.2.2. OpenRead

"OpenRead" refers to a concept or initiative related to making literature, particularly academic or scholarly works, freely accessible to readers without restrictions or paywalls. It embodies the principles of open access, which advocate for unrestricted online access to scholarly research for the purpose of promoting knowledge dissemination and collaboration.

Key features of OpenRead may include:

Free Access: OpenRead platforms provide unrestricted access to a wide range of academic literature, including research papers, journal articles, books, and other scholarly works.

Open Licensing: Content available on OpenRead platforms may be published under open licenses, such as Creative Commons licenses, which allow users to freely share, reuse, and adapt the content with proper attribution.

Community Engagement: OpenRead initiatives often involve collaboration with academic institutions, libraries, publishers, and funding agencies to promote open access principles and support sustainable models for publishing and dissemination.

Global Reach: By removing barriers to access, OpenRead platforms aim to reach a global audience of researchers, students, educators, policymakers, and the general public, fostering a more inclusive and equitable knowledge-sharing environment.

Search and Discovery Tools: OpenRead platforms may offer advanced search and discovery tools, metadata enrichment, and recommendation systems to help users find relevant content and navigate the vast array of available literature.

Support for Open Science: OpenRead initiatives align with the principles of open science by promoting transparency, reproducibility, and collaboration in research. They encourage researchers to openly share their findings, data, and methodologies to advance scientific knowledge and innovation.

19.2.3. ELICIT.ORG

"ELICIT" is a cutting-edge AI-powered research assistant designed to streamline and enhance various aspects of researchers' workflows. Leveraging advanced language models like GPT-3, ELICIT automates key tasks such as literature reviews, filtering study types, and facilitating the research flow. Here's an overview of its features and functionalities:

Literature Review Automation: ELICIT's primary workflow revolves around automating literature reviews. By simply posing a research question, users can access relevant papers and summaries of key information in an easy-to-navigate table format.

Comprehensive Paper Summaries: ELICIT provides summaries of the top papers related to a given query, including titles, abstracts, citations, DOIs, and PDF links. This allows researchers to quickly assess the relevance and significance of each paper.

Detailed Table View: In addition to paper summaries, ELICIT.ORG offers a table view displaying key information such as abstracts, interventions, outcomes measured, and the number of participants for each study.

Access to Relevant Studies and Citations: Users can explore relevant studies and citations related to their research query, providing additional context and supporting evidence.

Paper Information Search: ELICIT enables users to search for metadata about sources, population demographics, intervention details, study results, and methodology. This comprehensive information enhances the understanding of each paper's context and findings.

Export Options: ELICIT offers various export options, allowing users to save results in formats such as BIB or CSV for further analysis and reference.

Concept Search and Data Extraction: Users can conduct concept searches to explore related topics and extract data from PDF files by uploading them to the platform.

Citation Guidance: ELICIT provides guidance on how to cite sources obtained through the platform, including the elicit.org URL in citations to acknowledge the use of the AI research assistant.

19.2.4. CONSENSUS.APP

consensus.app is an advanced academic search engine that harnesses the power of artificial intelligence (AI) to extract insights from vast repositories of research papers. Here's an overview of its features and functionalities:

Academic Search Engine: consensus.app offers a comprehensive academic search experience, enabling users to input research questions and receive relevant answers extracted from over 200 million scholarly documents. Covering diverse fields from medical sciences to social sciences and economics, the platform provides access to peer-reviewed literature across a wide range of subjects.

Extracted and Aggregated Findings: The platform aggregates findings from relevant papers, presenting users with synthesized summaries of key insights and methodological details such as aims, design, participants, and findings. Users can access both summaries and full-text articles to delve deeper into the research.

Search Functionality: consensus.app supports various search types, including direct questions, relationships between two concepts, and open-ended concepts. Users can input queries such as "Does the death penalty reduce crime?" or explore the relationship between concepts like "Fish oil and depression."

Synthesize (Beta) / Consensus Meter: The platform's Synthesize feature, currently in beta, leverages AI to analyze a selection of studies and provide summaries along with a Consensus Meter illustrating their collective agreement. For example, a search for "Is white rice linked to diabetes?" may reveal outcomes indicating the percentage of studies suggesting a positive association, possible connection, or no link.

Consensus Plugin within ChatGPT: Consensus.app offers a plugin within ChatGPT, allowing users to access scientific research directly within the interface. This integration enables users to answer questions, find papers, and create content seamlessly within the ChatGPT environment.

19.2.5. SCITE.AI

scite.ai is a cutting-edge platform that offers Citations in Context, revolutionizing the way researchers access and interpret citation data. Here's an overview of its features and functionalities:

Citations in Context: scite.ai provides users with access to over 1.2 billion Citation Statements and metadata sourced from more than 195 million papers. By integrating with various research solutions, scite.ai offers a comprehensive and intuitive platform for exploring citation data.

Deep Learning Model: scite.ai utilizes a powerful deep learning model to analyze full-text articles and extract valuable insights about citations. For a given publication, users can discover:

- The number of times it has been cited by others.

- The context in which it was cited, with text excerpts displaying the citation location within each citing paper.

- An assessment of whether each citation provides supporting or contrasting evidence for the claims made in the publication, or simply mentions it.

Citation Visualization: scite.ai enables users to visualize all citations and citation statements associated with a specific publication in one place. This comprehensive view allows researchers to gain a deeper understanding of the scholarly impact and reception of their work.

Smart Citation Index: scite.ai acts as a smart citation index, displaying the context of citations and classifying their intent using advanced deep learning techniques. This classification helps researchers discern the significance and relevance of each citation in relation to the cited publication.

19.2.6. CONNECTED PAPERS

Connected Papers is a powerful tool for discovering related studies and visualizing connections between academic papers. Leveraging visual graphs and the Semantic Scholar Paper Corpus, which contains hundreds of millions of published papers across various scientific and social science fields, Connected Papers helps users navigate the scholarly landscape with ease. Here's an overview of its features and functionalities:

Related Studies Visualization: Connected Papers utilizes visual graphs or other intuitive methods to display relevant studies connected to a given paper or topic. Users can explore interconnected research papers, facilitating discovery and exploration within their field of interest.

Access to Semantic Scholar Paper Corpus: The platform is connected to the Semantic Scholar Paper Corpus, providing access to a vast repository of academic papers. Users can leverage this extensive database to explore a wide range of scientific and social science literature.

Origin Paper and Connected Papers: Users can select a paper, such as "SERVQUAL," to view the origin paper along with connected papers linked to major sources. These connected papers may include links to PDFs, DOIs, publisher sites, Semantic Scholar, and Google Scholar, enabling users to access additional information and resources.

Graph Visualization: Connected Papers generates visual graphs illustrating the origin paper and its connected papers, offering a visual representation of the scholarly network surrounding a specific topic or research area.

Links to Prior Works and Derivative Works: Users can explore links to prior works and derivative works related to the selected paper. Detailed citations provided by Semantic Scholar offer insights into the origin paper's impact and influence within the academic community.

Search Functionality: Users can search for related studies by work title or enter keywords about a specific topic, facilitating targeted exploration and discovery.

Download and Save: Connected Papers allows users to download their saved items in Bib format, providing a convenient way to organize and reference relevant academic literature.

19.2.7. PAPER DIGEST

Paper Digest is a comprehensive knowledge graph and natural language processing platform tailored for the technology domain, covering a wide range of areas including technology, biology/health, all science areas, business, humanities/social sciences, patents, and grants. Here's an overview of its features and functionalities:

Summary & Synthesis: Paper Digest provides summaries and syntheses of academic papers, offering users a concise overview of the key insights and findings. Users can access digests or summaries of papers related to their research interests, facilitating efficient knowledge acquisition and understanding.

Literature Review / Systematic Review: Paper Digest offers dedicated sections for literature reviews and systematic reviews, allowing users to explore curated collections of scholarly literature in various domains. These sections provide valuable resources for researchers conducting literature reviews or seeking comprehensive overviews of specific topics.

Search Console: The platform's search console enables users to conduct targeted searches across its vast repository of academic papers. Users can search for papers, literature reviews, or systematic reviews using keywords or specific criteria, facilitating efficient information retrieval and exploration.

Conference Digest: Paper Digest offers digests of conference papers, including those from prominent conferences such as the NIPS conference. Users can access curated collections of conference papers, making it easier to stay updated on the latest research trends and developments in their field.

Tech AI Tools and Expert AI Tools: Paper Digest provides a range of AI-powered tools tailored for researchers and professionals. These include literature review tools, literature search tools, question answering tools, text summarization tools, expert search tools, executive search tools, reviewer search tools, and patent lawyer search tools. These tools leverage AI technology to enhance research efficiency and productivity.

Digest Categories: Users can subscribe to various digest categories, including daily paper digests, conference papers digests, best paper digests, and topic tracking digests. By entering their areas of interest in their account settings, users can receive customized digests based on their preferences, ensuring they stay informed about the latest research developments relevant to their field.

19.2.8. RESEARCH RABBIT

Research Rabbit is a powerful platform offering citation-based mapping of academic articles, facilitating exploration of similar, early, and later works related to a given topic. Here's an overview of its features and functionalities:

Citation-Based Mapping: Research Rabbit leverages citation data to map relationships between academic articles, allowing users to explore similar, early, and later works related to their research interests. By analyzing citation patterns, the platform provides insights into the scholarly landscape and facilitates discovery of relevant literature.

Vast Coverage of Academic Materials: Research Rabbit boasts access to hundreds of millions of academic articles, covering more than 90% of materials found in major databases used by academic institutions, such as Scopus, Web of Science, and others. This extensive coverage ensures that users have access to a comprehensive repository of scholarly literature across various disciplines.

Search Algorithms: The platform employs advanced search algorithms borrowed from reputable sources such as the National

Institutes of Health (NIH) and Semantic Scholar. These algorithms enhance search accuracy and efficiency, enabling users to find relevant articles more effectively.

Untitled Collection and Recommendations: Research Rabbit automatically collects users' search histories in a default "Untitled Collection." Based on these search histories, the platform generates recommendations for three types of related results: Similar Works, Earlier Works, and Later Works. Users can visualize these recommendations in various formats, including network graphs, timelines, and lists of first authors.

Zotero Integration: Research Rabbit seamlessly integrates with Zotero, a popular reference management tool. Users can easily import and export data between Research Rabbit and Zotero, streamlining the citation management process and enhancing research workflow efficiency.

Export Options: Findings and results obtained through Research Rabbit can be exported in various formats, including BibTeX, RIS, and CSV. This flexibility allows users to save and organize their research findings according to their preferences and needs.

19.2.9. ChatGPT

GPT3.5 by OpenAI. Knowledge cutoff date is September 2021. According to ChatGPT, the generated responses are non-deterministic by default. So, if we run the searches again and get slightly or very different results, it's likely due to this factor. ChatGPT may find non-existent references, also.

According to an unpublished article "ChatGPT cites the most-cited articles and journals, relying solely on Google Scholar's citation counts"

How a search can be conducted through **ChatGPT is depicted as follows**

Examples - RESEARCH QUESTION

How HIV transmission occurs mothers to child?
During Pregnancy (In Utero Transmission)
During Childbirth (Intrapartum Transmission)
During Breastfeeding (Postnatal Transmission)

Example – SYNONYMS
What are the synonyms of mothers to child transmission?
Vertical transmission
Perinatal transmission
Maternal-infant transmission
Mother-to-infant transmission
Maternal-fetal transmission
Maternal-neonatal transmission
Intrauterine transmission
Prenatal transmission
Antenatal transmission
Intrapartum transmission
Postnatal transmission

Example - RELATED WORDS
What are the related words **of** mothers to child transmission?

Prevention of mother-to-child transmission (PMTCT), Vertical transmission, Perinatal HIV transmission, Maternal-fetal transmission, Antenatal transmission, Intrapartum transmission, Postnatal transmission, Maternal health, Pediatric HIV, Infant HIV, Maternal HIV infection, HIV-exposed infant, HIV-positive mother, Breastfeeding transmission, Early infant diagnosis (EID), Antiretroviral therapy (ART) for pregnant women, Infant prophylaxis, Early infant treatment, HIV testing during pregnancy, Maternal viral load suppression

Example of Searching with SUBJECT TERMS & KEYWORDS

Show me a search string that combines the use of PubMed MeSH terms and keywords on the mothers to child transmission of HIV.

(("HIV Infections/transmission"[Mesh] OR "HIV"[Mesh]) AND ("Vertical Transmission"[Mesh] OR "Perinatal Transmission"[Mesh] OR "Pregnancy Complications, Infectious"[Mesh] OR "Pregnancy Complications, Infectious/prevention and control"[Mesh] OR "Pregnancy Complications, Infectious/therapy"[Mesh] OR "Antiretroviral Therapy, Highly Active"[Mesh] OR "Prevention of Mother-to-Child Transmission"[Mesh] OR "Infant, Newborn"[Mesh]) AND (transmission OR MTCT OR PMTCT OR vertical transmission OR perinatal transmission OR maternal-fetal transmission OR antenatal transmission OR intrapartum transmission OR postnatal transmission OR maternal health OR pediatric HIV OR infant HIV OR maternal HIV infection

235

OR breastfeeding transmission OR early infant diagnosis OR antiretroviral therapy OR infant prophylaxis OR early infant treatment OR HIV testing during pregnancy OR maternal viral load suppression))

Comments and Suggestions:

#1 -This search string may not work on all database platforms. For example, Medline (OVID) may not be able to run it, Medline & CINAHL (EBSCO) can run it without any problems while PubMed will run it but it will indicate which MeSH terms are not indexed. See PubMed's search History and Details under Advanced.

#2 -This is just one possible combination. You may need to adjust it based on your task and the databases you are using.

#3 - It might be easier if you break the task into 3 parts: ask for synonyms > ask for thesauri terms > and ask it to combine them.

Examples   -   DATABASE RECOMMENDATIONS

Which databases to use to find articles on HOW HIV TRANSMISSION OCCURS MOTHERS TO CHILD?

1. PubMed: PubMed is a comprehensive database of biomedical literature from the National Library of Medicine. It covers a wide range of health-related topics, including HIV/AIDS research.

2. Embase: Embase is a biomedical and pharmacological database that provides

extensive coverage of biomedical literature, including articles related to HIV transmission from mothers to children.

3. Cochrane Library: The Cochrane Library is a collection of databases that provides systematic reviews and meta-analyses of healthcare interventions. It includes studies on the prevention and management of HIV transmission from mothers to children.

4. Scopus: Scopus is a multidisciplinary database that covers research from various fields, including health sciences. It includes articles from scientific journals, books, and conference proceedings related to HIV/AIDS research.

5. Web of Science: Web of Science is a multidisciplinary database that covers research from a wide range of fields, including health sciences. It includes articles from scientific journals, books, and conference proceedings relevant to HIV transmission.

6. CINAHL: CINAHL is a database of nursing and allied health literature. It includes articles on maternal and child health, as well as HIV/AIDS research related to mother-to-child transmission.

[Note: The response above was copied and pasted directly from ChatGPT. No editing was added.]

Example - GREY LITERATURE

Please show me all the resources that offer grey literature for conducting research in mother-to-child transmission of HIV.

1. ERIC (Education Resources Information Center): ERIC provides access to education research and information, including grey literature such as reports, conference papers, and working papers related to HIV/AIDS education, including mother-to-child transmission.

2. OpenGrey: OpenGrey is a database that provides access to grey literature produced in Europe. It includes reports, theses, conference papers, and other types of grey literature on various topics, including HIV/AIDS and mother-to-child transmission.

3. WorldCat: WorldCat is a global catalog of library collections. It includes records of books, articles, and other materials held by libraries worldwide, including grey literature related to HIV/AIDS and mother-to-child transmission.

4. ProQuest Dissertations & Theses Global: ProQuest Dissertations & Theses Global is a database that provides access to graduate-level research, including theses and dissertations related to HIV/AIDS research, including studies on mother-to-child transmission.

5. EThOS (Electronic Theses Online Service): EThOS is a database of theses

produced by UK higher education institutions. It includes electronic copies of theses related to HIV/AIDS research, including studies on mother-to-child transmission.

6. Institutional Repositories of universities and academic institutions: Many universities and academic institutions have institutional repositories where they archive and provide access to research outputs produced by their faculty, staff, and students. These repositories may include grey literature such as reports, working papers, and conference papers on HIV/AIDS and mother-to-child transmission.

7. AIDSinfo: AIDSinfo is a service of the U.S. Department of Health and Human Services that provides access to information on HIV/AIDS treatment, prevention, and research. It includes resources such as guidelines, fact sheets, and reports on mother-to-child transmission of HIV.

8. UNAIDS: UNAIDS is the joint United Nations program on HIV/AIDS. It provides access to reports, publications, and other resources related to HIV/AIDS research and programming, including mother-to-child transmission.

9. PubMed Central (PMC): PubMed Central is a free full-text archive of biomedical and life sciences journal literature. It includes articles from journals that deposit their content in PMC, including

grey literature related to HIV/AIDS research and mother-to-child transmission.

19.2.10. Nested Knowledge

Nested Knowledge offers a comprehensive software platform for systematic literature review and meta-analysis. The software is composed of two parts which work in tandem. Search, screen, extract data, and complete critical appraisal with AutoLit. Visualize, analyze, publish and share insights with Synthesis.

19.2.11. Litstream.

Litstream allows researchers to collaborate on literature screening and data extraction through an efficient platform that tracks the literature process meticulously and across multiple team members working simultaneously. Managers can design fit-for-purpose review steps, make assignments, and monitor progress at a granular level. Litstream helps ensure transparent documentation of the review process while prioritizing the most relevant literature to your team of subject matter experts.

Q: What is a systematic review?

A: A systematic review is a comprehensive and structured review of existing research literature on a specific topic.

Q: What is/ the main goal of a systematic review?

A: The main goal of a systematic review is to provide a comprehensive summary of the available evidence on a particular research question.

Q: What is the difference between a systematic review and a traditional literature review?

A: A systematic review follows a rigorous and systematic methodology, while a traditional literature review may not have a defined process or criteria.

Q: What is a meta-analysis?

A: A meta-analysis is a statistical technique used in systematic reviews to combine and analyze data from multiple studies to draw more robust conclusions.

Q: Why is a systematic review considered a higher level of evidence in research?

A: Systematic reviews are considered a higher level of evidence because they use a rigorous methodology to minimize bias and provide a comprehensive overview of existing research.

Q: What is the role of a systematic review protocol?

A: A systematic review protocol outlines the methods and criteria that will be used in the review, ensuring transparency and minimizing bias.

Q: What are the key steps in conducting a systematic review?

A: The key steps include defining the research question, searching for relevant studies, screening and selecting studies, data extraction, quality assessment, and synthesis of results.

Q: What is the PRISMA statement, and why is it important in systematic reviews?

A: PRISMA (Preferred Reporting Items for Systematic Reviews and Meta-Analyses) is a set of guidelines for reporting systematic reviews, enhancing transparency and quality.

Q: What is publication bias, and how can it affect the results of a systematic review?

A: Publication bias is the tendency for studies with positive results to be published more frequently, potentially skewing the results of a systematic review.

Q: What is inclusion and exclusion criteria in a systematic review?

A: Inclusion criteria define the characteristics a study must have to be included in the review, while exclusion criteria specify reasons for study exclusion.

Q: How do you assess the quality of studies included in a systematic review?

A: Quality assessment can involve tools like the Cochrane Risk of Bias Tool, which evaluates study design, sample size, and other factors affecting study validity.

Q: What is a forest plot in the context of a systematic review?

A: A forest plot is a graphical representation of the results of individual studies included in a meta-analysis, showing effect sizes and confidence intervals.

Q: What is heterogeneity in the context of systematic reviews?

A: Heterogeneity refers to the variability among study results in a systematic review. It can be assessed using statistical tests like the I^2 statistic.

Q: What is a subgroup analysis in a systematic review?

A: A subgroup analysis is performed to explore variations in treatment effects within specific subgroups of the study population.

Q: What is a sensitivity analysis in a systematic review?

A: A sensitivity analysis is conducted to assess the impact of methodological choices or the inclusion/exclusion of studies on the overall results.

Q: What is a narrative synthesis in a systematic review?

A: A narrative synthesis is a textual summary of study findings in a systematic review when meta-analysis is not possible.

Q: What is the role of a systematic review in evidence-based medicine?

A: Systematic reviews provide a foundation for evidence-based medicine by summarizing the best available evidence to inform clinical decisions.

Q: What is a scoping review, and how does it differ from a systematic review?

A: A scoping review aims to map the existing literature on a broad topic, while a systematic review focuses on answering a specific research question.

Q: What are grey literature sources, and why are they important in systematic reviews?

A: Grey literature includes unpublished or non-peer-reviewed sources such as conference abstracts and government reports. They are important to minimize publication bias.

Q: What is the role of the Cochrane Collaboration in systematic reviews?

A: The Cochrane Collaboration is an international organization that produces high-quality systematic reviews and promotes evidence-based healthcare.

Q: How can conflicts of interest be addressed in a systematic review?

A: Conflicts of interest can be managed by disclosing potential conflicts, involving unbiased reviewers, and following transparent methodologies.

Q: Can a systematic review include qualitative research studies?

A: Yes, systematic reviews can include qualitative research studies, and there are specific methodologies for synthesizing qualitative data.

Q: What is a systematic review update, and why is it necessary?

A: A systematic review update involves periodically revisiting and updating a review to incorporate new evidence and maintain its relevance.

Q: What is the role of a systematic review in policy-making?

A: Systematic reviews can inform policy decisions by providing evidence on the effectiveness and safety of interventions or policies.

Q: How can you address language bias in a systematic review?

A: Language bias can be addressed by conducting searches in multiple languages, translating relevant studies, and involving reviewers proficient in those languages.

Q: What is the difference between a systematic review and a rapid review?

A: A rapid review is a streamlined version of a systematic review that is conducted more quickly, often with some methodological shortcuts.

Q: What is the role of the GRADE framework in systematic reviews?

A: The GRADE framework helps assess the quality of evidence and make recommendations based on the strength0 of that evidence.

Q: How do you handle missing data in a systematic review?

A: Handling missing data involves contacting study authors for additional information and conducting sensitivity analyses to assess the impact of missing data.

Q: What is the role of a librarian or information specialist in a systematic review?

A: Librarians or information specialists help design search strategies, conduct literature searches, and manage references in systematic reviews.

Q: What are the potential biases associated with industry-funded studies in systematic reviews?

A: Industry-funded studies may be more likely to show favorable results, leading to potential bias if they are overrepresented in a systematic review.

Q: How can you assess the risk of bias in non-randomized studies included in a systematic Review?

A: Tools like the ROBINS-I (Risk of Bias In Non-randomized Studies - of Interventions) tool can be used to assess the risk of bias in non-randomized studies.

Q: What is the role of stakeholder involvement in a systematic review?

A: Stakeholder involvement can help ensure that the research question, outcomes, and methods of a systematic review align with the needs and perspectives of relevant stakeholders.

Q: What is the role of a systematic review in identifying research gaps?

A: Systematic reviews can help identify areas where further research is needed by highlighting gaps in the existing evidence.

Q: What is the significance of the "funnel plot" in systematic reviews?

A: Funnel plots are used to visually assess the potential presence of publication bias in a meta-analysis.

Q: What is the role of a systematic review in a systematic review?

A: A systematic review may include both published and unpublished.

Q: What is the primary purpose of conducting a systematic review?

A: The primary purpose of a systematic review is to provide a comprehensive and unbiased synthesis of existing research evidence on a specific research question or topic.

Q: How is a systematic review different from a traditional literature review?

A: Unlike a traditional literature review, which may be less structured and selective, a systematic review follows a strict and predefined methodology, including transparent search strategies and inclusion criteria.

Q: What are the key characteristics of studies included in a systematic review?

A: Studies included in a systematic review should meet predetermined inclusion criteria, be relevant to the research question, and undergo a rigorous quality assessment.

Q: Why is transparency crucial in the methods of a systematic review?

A: Transparency in methods is essential to ensure the reproducibility of a systematic review and to allow others to assess the validity of the findings.

Q: What is meant by the term "meta-analysis" in the context of systematic reviews?

A: Meta-analysis is a statistical technique used in systematic reviews to combine and analyze data from multiple studies, providing a quantitative summary of the evidence.

Q: How is data extraction conducted in a systematic review?

A: Data extraction involves systematically collecting relevant information from each included study, such as study design, sample size, and key outcomes, using standardized forms.

Q: What is the role of a systematic review protocol, and why is it important?

A: A systematic review protocol is a detailed plan that outlines the research question, inclusion criteria, search strategy, and methods. It is crucial for maintaining methodological rigor and transparency.

Q: What are the potential sources of bias in a systematic review, and how are they addressed?

A: Sources of bias in a systematic review can include publication bias, selection bias, and reporting bias. They are addressed through rigorous search strategies, quality assessments, and sensitivity analyses.

Q: How does a systematic review contribute to evidence-based practice in healthcare?

A: Systematic reviews play a vital role in evidence-based practice by providing healthcare professionals with a reliable summary of the best available evidence, helping them make informed decisions.

Q: What are the limitations of a systematic review, and how can they be mitigated?

A: Limitations of systematic reviews can include the availability of limited data, heterogeneity among studies, and potential biases. These limitations can be mitigated through transparent reporting and cautious interpretation of findings.

List of some legendary books

1. "Systematic Reviews in Health Care: Meta-Analysis in Context" by Matthias Egger, George Davey Smith, and Douglas G. Altman (Publisher: Wiley-Blackwell).
2. "Cochrane Handbook for Systematic Reviews of Interventions" by Julian PT Higgins and Sally Green (Publisher: Wiley).
3. "Systematic Approaches to a Successful Literature Review" by Andrew Booth, Anthea Sutton, and Diana Papaioannou (Publisher: Sage Publications).
4. "Meta-Analysis and Systematic Reviews in the Social Sciences: A Practical Guide" by Mark W. Lipsey

and David B. Wilson (Publisher: Sage Publications).
5. "Systematic Reviews: Synthesis of Best Evidence for Healthcare Decisions" by Jos Kleijnen and Lotty Hooft (Publisher: Springer).
6. "A Beginner's Guide to Evidence-Based Practice in Health and Social Care" by Helen Aveyard (Publisher: McGraw-Hill Education).
7. "Introduction to Systematic Reviews" by David Gough, Sandy Oliver, and James Thomas (Publisher: Sage Publications).
8. "Meta-Analysis: A Structural Equation Modeling Approach" by Mike W.-L. Cheung (Publisher: Wiley).
9. "Systematic Reviews to Support Evidence-Based Medicine" by Khalid S. Khan, Regina Kunz, Jos Kleijnen, and Gerd Antes (Publisher: CRC Press).
10. "Systematic Reviews and Meta-Analysis: A Step-by-Step Guide" by H. K. Kang and S. M. W. Park (Publisher: Springer).
11. "Conducting Research Literature Reviews: From the Internet to Paper" by Arlene G. Fink (Publisher: Sage Publications).

12. "Systematic Reviews and Meta-Analysis" by Michael E. Hyland (Publisher: Oxford University Press).
13. "Systematic Review and Meta-Analysis: A Practical Guide" by Julia H. Littell (Publisher: Sage Publications).
14. "Critical Appraisal of Epidemiological Studies and Clinical Trials" by Mark Elwood (Publisher: Oxford University Press).
15. "Systematic Reviews in Educational Research: Methodology, Perspectives, and Application" by T. Plucker and M. Stocking (Publisher: Springer).
16. "Meta-Analysis: Methods, Strengths, Weaknesses, and Political Uses" by Gene V. Glass (Publisher: Waveland Press).
17. "Meta-Analysis: New Developments and Applications in Medical and Social Sciences" by Wolfgang Karl Härdle and Yarema Okhrin (Publisher: Springer).
18. "Evidence Synthesis for Decision Making in Healthcare" by Nicky J. Welton, Alex J. Sutton, Keith R. Abrams, and David R. Jones (Publisher: Wiley).
19. "Integrating Research: A Guide for Literature Reviews" by Harris M.

Cooper (Publisher: Sage Publications).
20. "Meta-Analysis: An Updated Collection from the Stata Journal" edited by Jonathan Sterne (Publisher: Stata Press).

- Please note that the availability of these books may vary depending on your location and the latest publications in the field. It's a good idea to check with your local libraries, online booksellers, or academic institutions to access these resources.

Matthias Egger

Matthias Egger is a professor of epidemiology and public health at the University of Bern, Switzerland. Since January 2017, Matthias Egger is president of the Research Council of the Swiss National Science Foundation. In 1997, Egger published a paper describing a method for detecting bias in meta-analyses by analyzing funnel plots. Till date this paper has been cited more than 48000 times. In 2005, Egger published a study comparing 110 trials of homeopathy with 110 trials of conventional medicine in the Lancet. It found that there was strong evidence that conventional medicine was more effective than placebo, but only weak evidence that homeopathy was. Egger has also published research on a wide variety of other medical topics, such as the demographics of people who choose assisted suicide, the association between exposure to aircraft noise and heart attacks, and the effectiveness of pneumococcal polysaccharide vaccines.

LEGEND

Other legends of Evidence Synthesis

1. Kay Dickersin: Kay Dickersin is a prominent researcher and advocate for improving the quality and transparency of systematic reviews. She co-founded the Cochrane Eye and Vision Group and has been instrumental in developing standards for conducting and reporting systematic reviews, particularly in the field of ophthalmology.
2. Julian Higgins: Julian Higgins is a leading methodologist in systematic review and meta-analysis. He has contributed to the development of statistical methods for synthesizing evidence from multiple studies, including meta-regression, network meta-analysis, and publication bias assessment.
3. Lisa Bero: Lisa Bero is a prominent researcher in evidence-based healthcare and systematic review methodology. Her work focuses on understanding and addressing biases in evidence synthesis, conflicts of interest, and the influence of industry funding on research outcomes.
4. Mike Clarke: Mike Clarke is a co-founder of the Cochrane Collaboration and has played a key role in promoting the use of systematic reviews to inform healthcare policy and practice. He has contributed to the development of guidelines and resources for conducting high-quality systematic reviews across various disciplines.
5. Paul Glasziou: Paul Glasziou is a clinical epidemiologist and advocate for evidence-based medicine. He has contributed to the

development of methods for synthesizing evidence and has emphasized the importance of applying evidence from systematic reviews in clinical decision-making.

6. Gordon Guyatt: Gordon Guyatt is a physician and researcher known for his contributions to evidence-based medicine and the development of the GRADE (Grading of Recommendations Assessment, Development, and Evaluation) approach. GRADE provides a systematic framework for assessing the quality of evidence and strength of recommendations in systematic reviews and clinical guidelines.

7. Prathap Tharyan: Prathap Tharyan is a psychiatrist and researcher known for his work in evidence-based medicine and systematic reviews. He has contributed to the development of systematic review methodology and guidelines, particularly in the field of mental health.

8. Sangeeta Desai: Sangeeta Desai is a public health researcher who has conducted systematic reviews and meta-analyses on various topics, including maternal and child health, nutrition, and infectious diseases. She has contributed to evidence synthesis efforts aimed at informing public health policy and practice in India.

9. Howard White: Howard White is a prominent figure in the field of evidence synthesis and international development. He has made significant contributions to the advancement of evidence-based policy and practice, particularly in the areas of impact evaluation, systematic reviews, and evidence synthesis.

Howard White is the former Executive Director of the Campbell Collaboration, an international research network that produces systematic reviews and evidence syntheses in the social, behavioral, and educational fields. Under his leadership, the Campbell Collaboration expanded its scope and impact, promoting the use of rigorous evidence synthesis methods to inform decision-making and policy formulation.

10. Lalit Dandona: Lalit Dandona is a public health researcher known for his work in epidemiology, health systems research, and global health. He has led systematic review projects on various topics, including HIV/AIDS, tuberculosis, and non-communicable diseases, with a focus on addressing health disparities and improving health outcomes in India.

Bibliography

- "A brief history of Cochrane". community.cochrane.org, 2019.
- "Animated Storyboard: What Are Systematic Reviews?". cccrg.cochrane.org. Cochrane Consumers and Communication.
- "Animated Storyboard: What Are Systematic Reviews?". cccrg.cochrane.org.
- "Cochrane crowd". crowd.cochrane.org.
- "Cochrane's work on Rapid Reviews in response to COVID-19". www.cochrane.org.
- "Environmental Evidence: Reliable evidence, informed decisions, better environment". www.environmentalevidence.org.
- "EPPI-Centre Home". eppi.ioe.ac.uk.
- "Half of all clinical trials have never reported results". AllTrials. 2015.
- "Handsearching". training.cochrane.org.
- "History of Systematic Reviews". Evidence for Policy and Practice Information and Co-ordinating Centre (EPPI-Centre).
- "History of the Campbell Collaboration". Campbell Collaboration.
- "Home". covidrapidreviews.cochrane.org.
- "How to disseminate your research". www.nihr.ac.uk.
- "Living systematic reviews". community.cochrane.org.

- "Main page". Cochrane Library.
- "MetaXL software page". Epigear.com. 3 2017.
- "MetaXL User Guide" (PDF).
- "Methodological Expectations of Cochrane Intervention Reviews (MECIR)". Cochrane.
- "Preferred Reporting Items for Systematic Reviews and Meta-Analyses: The PRISMA Statement". www.equator-network.org.
- "PRISMA". www.prisma-statement.org. 0000000
- "PROSPERO". Centre for Reviews and Dissemination. University of York.
- "Report on Certain Enteric Fever Inoculation Statistics". British Medical Journal. 2 (2). British Medical Journal Publishing Group: 1243–1246. (1904).
- "The Cochrane-Wikipedia partnership in 2016". Cochrane.
- "The difference we make". www.cochrane.org.
- "The PRISMA statement". Prisma-statement.org, 2012.
- "What is a rapid review? Systematic Review Library Guides at CQ University". library.cqu.edu.au.
- "What is EBM?". Centre for Evidence Based Medicine. 2009.
- "What Students Are Saying About ChatGPT". The New York Times. 2023.
- Adams, John; Khan, Hafiz T. A.; Raeside, Robert (2007). Research methods for graduate business and social science

students. New Delhi: SAGE Publications. p. 56. ISBN 9780761935896.
- Adèr HJ (2008). Advising on Research Methods: A Consultant's Companion. Johannes van Kessel Publishing. ISBN 978-90-79418-01-5.
- Ader HJ, Mellenbergh GJ, Hand DJ (2008). "Methodological quality". Advising on Research Methods: A consultant's companion. Johannes van Kessel Publishing. ISBN 978-90-79418-02-2.
- Akhigbe T, Zolnourian A, Bulters D (2017). "Compliance of systematic reviews articles in brain arteriovenous malformation with PRISMA statement guidelines: Review of literature". Journal of Clinical Neuroscience. 39: 45–48.
- Al Khalaf MM, Thalib L, Doi SA (2011). "Combining heterogenous studies using the random-effects model is a mistake and leads to inconclusive meta-analyses". Journal of Clinical Epidemiology. 64 (2): 119–123.
- Alkaissi, Hussam; McFarlane, Samy I.; Alkaissi, Hussam; McFarlane, Samy I. (2023). "Artificial Hallucinations in ChatGPT: Implications in Scientific Writing". Cureus. 15 (2): e35179.
- Altman DG (1994). "The scandal of poor medical research". BMJ. 308 (6924): 283–284.
- Arksey H, O'Malley L (2005). "Scoping studies: towards a methodological framework" (PDF). International Journal

of Social Research Methodology. 8 (1): 19–32.
- ❖ Arksey H, O'Malley L (2005). "Scoping studies: Towards a methodological framework" (PDF). International Journal of Social Research Methodology. 8: 19–32.
- ❖ Armstrong R, Hall BJ, Doyle J, Waters E "Cochrane Update. 'Scoping the scope' of a Cochrane Review". Journal of Public Health. 33 (1): 147–150.
- ❖ Baglione, L. (2012). Writing a Research Paper in Political Science. Thousand Oaks, California: CQ Press.
- ❖ Baker, P. (2000). "Writing a Literature Review". The Marketing Review. 1 (2): 219–247.
- ❖ Banno M, Tsujimoto Y, Luo Y, Miyakoshi C, Kataoka Y (2021). "CAST-HSROC: A Web Application for Calculating the Summary Points of Diagnostic Test Accuracy from the Hierarchical Summary Receiver Operating Characteristic Model". Cureus. 13 (2): e13257.
- ❖ Bartoš F, Maier M, Quintana D, Wagenmakers EJ (16 2020). "Adjusting for Publication Bias in JASP & R - Selection Models, PET-PEESE, and Robust Bayesian Meta-Analysis". Advances in Methods and Practices in Psychological Science.
- ❖ Bartoš F, Maier M, Wagenmakers EJ, Goosen J, Denwood M, Plummer M (20 2022). "RoBMA: An R Package for Robust

- Bayesian Meta-Analyses". Retrieved 9 2022.
- ❖ Bearman M, Dawson P (2013). "Qualitative synthesis and systematic review in health professions education". Medical Education. 47 (3): 252–260.
- ❖ Bhandari, Mohit; Devereaux, P. J.; Guyatt, Gordon H.; Cook, Deborah J.; Swiontkowski, Marc F.; Sprague, Sheila; Schemitsch, Emil H. (2002). "An Observational Study of Orthopaedic Abstracts and Subsequent Full-Text Publications". The Journal of Bone and Joint Surgery-American Volume. 84 (4): 615–621.
- ❖ Bilotta GS, Milner AM, Boyd I (2014). "On the use of systematic reviews to inform environmental policies". Environmental Science & Policy. 42: 67–77.
- ❖ Bolderston, Amanda (2008). "Writing an Effective Literature Review". Journal of Medical Imaging and Radiation Sciences. 39 (2): 86–92.
- ❖ Booth A, Noyes J, Flemming K, Gerhardus A, Wahlster P, Van Der Wilt GJ, et al. (2016). Guidance on choosing qualitative evidence synthesis methods for use in health technology assessments of complex interventions. p. 32. OCLC 944453327.
- ❖ Briscoe S (2018). "A review of the reporting of web searching to identify studies for Cochrane systematic reviews". Research Synthesis Methods. 9 (1): 89–99.

- Brockwell SE, Gordon IR (2001). "A comparison of statistical methods for meta-analysis". Statistics in Medicine. 20 (6): 825–840.
- Brockwell SE, Gordon IR (2007). "A simple method for inference on an overall effect in meta-analysis". Statistics in Medicine. 26 (25): 4531–4543.
- Brooks SP, Gelman A (1998). "General methods for monitoring convergence of iterative simulations" (PDF). Journal of Computational and Graphical Statistics. 7 (4): 434–455.
- Bucher HC, Guyatt GH, Griffith LE, Walter SD (1997). "The results of direct and indirect treatment comparisons in meta-analysis of randomized controlled trials". Journal of Clinical Epidemiology. 50 (6): 683–691.
- Burke DL, Ensor J, Riley RD (2017). "Meta-analysis using individual participant data: one-stage and two-stage approaches, and why they differ". Statistics in Medicine. 36 (5): 855–875.
- Cartabellotta A, Tilson JK (2019). "The ecosystem of evidence cannot thrive without efficiency of knowledge generation, synthesis, and translation". Journal of Clinical Epidemiology. 110: 90–95.
- Clarke, Mike; Chalmers, Iain (2018). "Reflections on the history of systematic reviews". BMJ Evidence-Based Medicine. 23 (4): 121–122.

- Cliche, Mathieu; Rosenberg, David; Madeka, Dhruv; Yee, Connie (2017), Ceci, Michelangelo; Hollmén, Jaakko; Todorovski, Ljupčo; Vens, Celine (eds.), "Scatteract: Automated Extraction of Data from Scatter Plots", Machine Learning and Knowledge Discovery in Databases, vol. 10534, Cham: Springer International Publishing, pp. 135–150,
- Cochrane AL (1972). Effectiveness and efficiency: random reflections on health services (PDF). [London]: Nuffield Provincial Hospitals Trust. ISBN 0-900574-17-8. OCLC 741462.
- Colquhoun HL, Levac D, O'Brien KK, Straus S, Tricco AC, Perrier L, et al. (2014). "Scoping reviews: time for clarity in definition, methods, and reporting". Journal of Clinical Epidemiology. 67 (12): 1291–1294.
- Conn, Vicki S.; Valentine, Jeffrey C.; Cooper, Harris M.; Rantz, Marilyn J. (2003). "Grey Literature in Meta-Analyses". Nursing Research. 52 (4): 256–261.
- Coon JT, Orr N, Shaw L, Hunt H, Garside R, Nunns M, et al. (2022). "Bursting out of our bubble: using creative techniques to communicate within the systematic review process and beyond". Systematic Reviews. 11 (1): 56.
- Cowie K, Rahmatullah A, Hardy N, Holub K, Kallmes K (2022). "Web-Based Software Tools for Systematic Literature Review in Medicine: Systematic Search

and Feature Analysis". JMIR. 10 (5): e33219.
- Cramer, Duncan (2003). "A Cautionary Tale of Two Statistics: Partial Correlation and Standardized Partial Regression". The Journal of Psychology. 137 (5): 507–511.
- Debray TP, Moons KG, Abo-Zaid GM, Koffijberg H, Riley RD (2013). "Individual participant data meta-analysis for a binary outcome: one-stage or two-stage?". PLOS ONE. 8 (4): e60650.
- Debray TP, Moons KG, van Valkenhoef G, Efthimiou O, Hummel N, Groenwold RH, Reitsma JB (2015). "Get real in individual participant data (IPD) meta-analysis: a review of the methodology". Research Synthesis Methods. 6 (4): 293–309.
- Doi SA, Barendregt JJ, Khan S, Thalib L, Williams GM (2015). "Simulation Comparison of the Quality Effects and Random Effects Methods of Meta-analysis". Epidemiology. 26 (4): e42–e44.
- Doi SA, Barendregt JJ, Khan S, Thalib L, Williams GM (2015). "Advances in the meta-analysis of heterogeneous clinical trials I: The inverse variance heterogeneity model". Contemporary Clinical Trials. 45 (Pt A): 130–138.
- Doi SA, Barendregt JJ, Khan S, Thalib L, Williams GM (2015). "Advances in the meta-analysis of heterogeneous clinical trials II: The quality effects model". Contemporary Clinical Trials. 45 (Pt A): 123–129.

- Doi SA, Barendregt JJ, Mozurkewich EL (2011). "Meta-analysis of heterogeneous clinical trials: an empirical example". Contemporary Clinical Trials. 32 (2): 288–298.
- Doi SA, Thalib L (2008). "A quality-effects model for meta-analysis". Epidemiology. 19 (1): 94–100.
- Durach CF, Kembro J, Wieland A (2017). "A New Paradigm for Systematic Literature Reviews in Supply Chain Management". Journal of Supply Chain Management. 53 (4): 67–85.
- E JY, Saldanha IJ, Canner J, Schmid CH, Le JT, Li T (2020). "Adjudication rather than experience of data abstraction matters more in reducing errors in abstracting data in systematic reviews". Research Synthesis Methods. 11 (3): 354–362.
- Egger, M; Jüni, P; Bartlett, C; Holenstein, F; Sterne, J (2003). "How important are comprehensive literature searches and the assessment of trial quality in systematic reviews? Empirical study". Health Technology Assessment. 7 (1): 1–82.
- Elliott JH, Turner T, Clavisi O, Thomas J, Higgins JP, Mavergames C, Gruen RL (2014). "Living systematic reviews: an emerging opportunity to narrow the evidence-practice gap". PLOS Medicine. 11 (2): e1001603.
- Evans I, Thornton H, Chalmers I, Glasziou P (2011). Testing treatments: better

- research for better healthcare (Second ed.). London: Pinter & Martin. ISBN 978-1-905177-48-6. OCLC 778837501.
- Eysenck, H. J. (1978). "An exercise in mega-silliness". American Psychologist. 33 (5): 517.
- Eysenck, H.J. (1995). "Meta-analysis or best-evidence synthesis?". Journal of Evaluation in Clinical Practice. 1 (1): 29–36.
- Ferguson (2015). "Retraction and republication for Lancet Resp Med tracheostomy paper". Retraction Watch.
- Ferguson C (2015). "BioMed Central retracting 43 papers for fake peer review". Retraction Watch.
- Ferguson CJ, Brannick MT (2012). "Publication bias in psychological science: prevalence, methods for identifying and controlling, and implications for the use of meta-analyses". Psychological Methods. 17 (1): 120–128
- Field, Andy P.; Gillett, Raphael (2010). "How to do a meta-analysis". British Journal of Mathematical and Statistical Psychology. 63 (3): 665–694.
- Flemming K, Booth A, Hannes K, Cargo M, Noyes J (2018). "Cochrane Qualitative and Implementation Methods Group guidance series-paper 6: reporting guidelines for qualitative, implementation, and process evaluation evidence syntheses" (PDF). Journal of Clinical Epidemiology. 97: 79–85.

- Flinders University Library. "Search Smart: Systematic Reviews: Methodology overview". flinders.libguides.com.
- Freeman SC, Kerby CR, Patel A, Cooper NJ, Quinn T, Sutton AJ (2019). "Development of an interactive web-based tool to conduct and interrogate meta-analysis of diagnostic test accuracy studies: MetaDTA". BMC Medical Research Methodology. 19 (1): 81.
- Gates, Simon; Ealing, Elizabeth (2019). "Reporting and interpretation of results from clinical trials that did not claim a treatment difference: survey of four general medical journals". BMJ Open. 9 (9): e024785.
- Glass GV, McGaw B, Smith ML (1981). Meta-analysis in social research. Beverly Hills, California: Sage Publications. ISBN 978-0-8039-1633-3.
- Glass GV, Smith ML, et al. (Far West Lab. for Educational Research and Development, San Francisco, CA) (1978). Meta-Analysis of Research on the Relationship of Class-Size and Achievement. The Class Size and Instruction Project. Washington, D.C.]: Distributed by ERIC Clearinghouse.
- Golder S, Loke Y, McIntosh HM (2008). "Poor reporting and inadequate searches were apparent in systematic reviews of adverse effects". Journal of Clinical Epidemiology. 61 (5): 440–448.

- Gough D, Oliver S, Thomas J (2017). An Introduction to Systematic Reviews (2nd ed.). London: Sage. p. XIII.
- Grames, Eliza M.; Stillman, Andrew N.; Tingley, Morgan W.; Elphick, Chris S. (2019). Freckleton, Robert (ed.). "An automated approach to identifying search terms for systematic reviews using keyword co-occurrence networks". Methods in Ecology and Evolution. 10 (10): 1645–1654.
- Granello, D. H. (2001). "Promoting cognitive complexity in graduate written work: Using Bloom's taxonomy as a pedagogical tool to improve Literature Reviews". Counselor Education & Supervision. 40 (4): 292–307.
- Grant MJ, Booth A (2009). "A typology of reviews: an analysis of 14 review types and associated methodologies". Health Information and Libraries Journal. 26 (2): 91–108.
- Gronau QF, Heck DW, Berkhout SW, Haaf JM, Wagenmakers EJ (2021). "A Primer on Bayesian Model-Averaged Meta-Analysis". Advances in Methods and Practices in Psychological Science. 4 (3).
- Gross, Arnd; Schirm, Sibylle; Scholz, Markus (2014). "Ycasd– a tool for capturing and scaling data from graphical representations". BMC Bioinformatics. 15 (1): 219.
- Haddaway NR, Bannach-Brown A, Grainger MJ, Hamilton WK, Hennessy

EA, Keenan C, et al. (2022). "The evidence synthesis and meta-analysis in R conference (ESMARConf): levelling the playing field of conference accessibility and equitability". Systematic Reviews. 11 (1): 113
- ❖ Haddaway, N.R.; Woodcock, P.; Macura, B.; Collins, A. (2015). "Making literature reviews more reliable through application of lessons from systematic reviews". Conservation Biology. 29 (6): 1596–1605.
- ❖ Haddaway, Neal R.; Macura, Biljana; Whaley, Paul; Pullin, Andrew S. (2018). "ROSES RepOrting standards for Systematic Evidence Syntheses: pro forma, flow-diagram and descriptive summary of the plan and conduct of environmental systematic reviews and systematic maps". Environmental Evidence. 7 (1).
- ❖ Hagen-Zanker J, Duvendack M, Mallett R, Slater R, Carpenter S, Tromme M (2012). "Making systematic reviews work for international development research". Overseas Development Institute.
- ❖ Haman, Michael; Školník, Milan (2023). "Using ChatGPT to conduct a literature review". Accountability in Research: 1–3.
- ❖ Harden A, Thomas J, Cargo M, Harris J, Pantoja T, Flemming K, et al. (2018). "Cochrane Qualitative and Implementation Methods Group guidance series-paper 5: methods for integrating qualitative and

implementation evidence within intervention effectiveness reviews" (PDF). Journal of Clinical Epidemiology. 97: 70-78.
- Hartling, Lisa; Featherstone, Robin; Nuspl, Megan; Shave, Kassi; Dryden, Donna M.; Vandermeer, Ben (2017). "Grey literature in systematic reviews: a cross-sectional study of the contribution of non-English reports, unpublished studies and dissertations to the results of meta-analyses in child-relevant reviews". BMC Medical Research Methodology. 17 (1): 64.
- Heck DW, Gronau QF, Wagenmakers EJ, Patil I (2021). "metaBMA: Bayesian model averaging for random and fixed effects meta-analysis". CRAN.
- Hedges LV, Vevea JL (1996). "Estimating Effect Size Under Publication Bias: Small Sample Properties and Robustness of a Random Effects Selection Model". Journal of Educational and Behavioral Statistics. 21 (4): 299-332.
- Hedges, Larry V.; Vevea, Jack L. (1998). "Fixed- and random-effects models in meta-analysis". Psychological Methods. 3 (4): 486-504.
- Helfenstein U (2002). "Data and models determine treatment proposals--an illustration from meta-analysis". Postgraduate Medical Journal. 78 (917): 131-134.
- Higgins JP, Altman DG, Gøtzsche PC, Jüni P, Moher D, Oxman AD, et al. (2011).

"The Cochrane Collaboration's tool for assessing risk of bias in randomised trials". BMJ. 343: d5928.
- ❖ Higgins JP, Green S (eds.). "Cochrane handbook for systematic reviews of interventions, version 5.1.0 (updated 2011)". The Cochrane Collaboration.
- ❖ Higgins JP, López-López JA, Becker BJ, Davies SR, Dawson S, Grimshaw JM, et al. (2019). "Synthesising quantitative evidence in systematic reviews of complex health interventions". BMJ
- ❖ Higgins JP, Thomas J, Chandler J, Cumpston M, Li T, Page MJ, Welch VA, eds. (2019). Cochrane Handbook for Systematic Reviews of Interventions. version 6.1. pp. section 4.6.
- ❖ Higgins JP, Thomas J, Chandler J, Cumpston M, Li T, Page MJ, Welch VA, eds. (2019). "Chapter 4: Searching for and selecting studies". Cochrane Handbook for Systematic Reviews of Interventions. version 6.1. section 4.6.
- ❖ Hopewell, Sally; Clarke, Mike (2005). "Abstracts presented at the American Society of Clinical Oncology conference: how completely are trials reported?". Clinical Trials. 2 (3): 265–268.
- ❖ Hunter JE, Schmidt FL (1990). Methods of Meta-Analysis: Correcting Error and Bias in Research Findings. Newbury Park, California; London; New Delhi: SAGE Publications.
- ❖ Hunter JE, Schmidt FL, Jackson GB, et al. (American Psychological Association.

- Division of Industrial-Organizational Psychology) (1982). Meta-analysis: cumulating research findings across studies. Beverly Hills, California: Sage. ISBN 978-0-8039-1864-1.
- Iacobucci G (2016). "Nearly half of all trials run by major sponsors in past decade are unpublished". BMJ. 355: i5955.
- Ioannidis JP, Trikalinos TA (2007). "The appropriateness of asymmetry tests for publication bias in meta-analyses: a large survey". CMAJ. 176 (8): 1091–1096.
- Ioannidis, John P.A. (2016). "The Mass Production of Redundant, Misleading, and Conflicted Systematic Reviews and Meta-analyses". The Milbank Quarterly. 94 (3): 485–514.
- Jackson D, Bowden J (2009). "A re-evaluation of the 'quantile approximation method' for random effects meta-analysis". Statistics in Medicine. 28 (2): 338–348.
- Koffel JB (2015). "Use of recommended search strategies in systematic reviews and the impact of librarian involvement: a cross-sectional survey of recent authors". PLOS ONE. 10 (5): e0125931.
- Koffel JB, Rethlefsen ML (2016). Thombs BD (ed.). "Reproducibility of Search Strategies Is Poor in Systematic Reviews Published in High-Impact Pediatrics, Cardiology and Surgery Journals: A Cross-Sectional Study". PLOS ONE. 11 (9): e0163309.

- Kontopantelis E, Reeves D (2010). "Metaan: Random-effects meta-analysis". Stata Journal. 10 (3): 395–407.
- Kontopantelis E, Reeves D (2013). "A short guide and a forest plot command (ipdforest) for one-stage meta-analysis". Stata Journal. 13 (3): 574–587.
- Kontopantelis E, Springate DA, Reeves D (2013). Friede T (ed.). "A re-analysis of the Cochrane Library data: the dangers of unobserved heterogeneity in meta-analyses". PLOS ONE. 8 (7): e69930.
- Kriston L (2013). "Dealing with clinical heterogeneity in meta-analysis. Assumptions, methods, interpretation". International Journal of Methods in Psychiatric Research. 22 (1): 1–15.
- Langan, Dean; Higgins, Julian P.T.; Jackson, Dan; Bowden, Jack; Veroniki, Areti Angeliki; Kontopantelis, Evangelos; Viechtbauer, Wolfgang; Simmonds, Mark (2019). "A comparison of heterogeneity variance estimators in simulated random-effects meta-analyses". Research Synthesis Methods. 10 (1): 83–98.
- Lau J, Antman EM, Jimenez-Silva J, Kupelnick B, Mosteller F, Chalmers TC (1992). "Cumulative meta-analysis of therapeutic trials for myocardial infarction". The New England Journal of Medicine. 327 (4): 248–254.
- LeBel E, Peters K (2011). "Compliance of Systematic Reviews in Plastic Surgery

with the PRISMA Statement". JAMA Facial Plastic Surgery. 18 (2): 101-105.

❖ Lefebvre, Carol; Glanville, Julie; Briscoe, Simon; Littlewood, Anne; Marshall, Chris; Metzendorf, Maria-Inti; Noel-Storr, Anna; Rader, Tamara; Shokraneh, Farhad (23 2019), Higgins, Julian P.T.; Thomas, James; Chandler, Jacqueline; Cumpston, Miranda (eds.), "Searching for and selecting studies", Cochrane Handbook for Systematic Reviews of Interventions (1 ed.), Wiley, pp. 67-107,

❖ LeLorier J, Grégoire G, Benhaddad A, Lapierre J, Derderian F (1997). "Discrepancies between meta-analyses and subsequent large randomized, controlled trials". The New England Journal of Medicine. 337 (8): 536-542.

❖ Levac D, Colquhoun H, O'Brien KK (2010). "Scoping studies: advancing the methodology". Implementation Science. 5 (1): 69.

❖ Li L, Tian J, Tian H, Moher D, Liang F, Jiang T, et al. (2014). "Network meta-analyses could be improved by searching more sources and by involving a librarian". Journal of Clinical Epidemiology. 67 (9): 1001-1007.

❖ Light RJ, Pillemer DB (1984). Summing up: the science of reviewing research. Cambridge, Massachusetts: Harvard University Press. ISBN 978-0-674-85431-4.

- Linnenluecke, Martina K; Marrone, Mauricio; Singh, Abhay K (2020). "Conducting systematic literature reviews and bibliometric analyses". Australian Journal of Management. 45 (2): 175–194.
- Lundh A, Lexchin J, Mintzes B, Schroll JB, Bero L, et al. (Cochrane Methodology Review Group) (2017). "Industry sponsorship and research outcome". The Cochrane Database of Systematic Reviews. 2017 (2): MR000033.
- Marshall C, Brereton P (2015). "Systematic review toolbox". Proceedings of the 19th International Conference on Evaluation and Assessment in Software Engineering. EASE '15. Nanjing, China: Association for Computing Machinery. pp. 1–6.
- McAuley, Laura; Pham, Ba'; Tugwell, Peter; Moher, David (2000). "Does the inclusion of grey literature influence estimates of intervention effectiveness report in meta-analyses?". The Lancet. 356 (9237): 1228–1231.
- McGowan J, Sampson M, Salzwedel DM, Cogo E, Foerster V, Lefebvre C (2016). "PRESS Peer Review of Electronic Search Strategies: 2015 Guideline Statement". Journal of Clinical Epidemiology. 75: 40–46.
- McGuinness, Luke A.; Higgins, Julian P. T. (2021). "Risk-of-bias VISualization (robvis): An R package and Shiny web app for visualizing risk-of-bias

- assessments". Research Synthesis Methods. 12 (1): 55–61.
- Meert D, Torabi N, Costella J (2016). "Impact of librarians on reporting of the literature searching component of pediatric systematic reviews". Journal of the Medical Library Association. 104 (4): 267–277.
- Moher D, Shamseer L, Clarke M, Ghersi D, Liberati A, Petticrew M, et al. (2015). "Preferred reporting items for systematic review and meta-analysis protocols (PRISMA-P) 2015 statement". Systematic Reviews. 4 (1): 1.
- Moher D, Tetzlaff J, Tricco AC, Sampson M, Altman DG (2007). "Epidemiology and reporting characteristics of systematic reviews". PLOS Medicine. 4 (3): e78.
- Moher, David; Tetzlaff, Jennifer; Tricco, Andrea C; Sampson, Margaret; Altman, Douglas G (2007). Clarke, Mike (ed.). "Epidemiology and Reporting Characteristics of Systematic Reviews". PLOS Medicine. 4 (3): e78.
- Moreau, David; Gamble, Beau (2022). "Conducting a meta-analysis in the age of open science: Tools, tips, and practical recommendations". Psychological Methods. 27 (3): 426–432.
- Mullins MM, DeLuca JB, Crepaz N, Lyles CM (2014). "Reporting quality of search methods in systematic reviews of HIV behavioral interventions (2000-2010): are the searches clearly explained, systematic and reproducible?".

Research Synthesis Methods. 5 (2): 116–130.
- Munn Z, Peters MD, Stern C, Tufanaru C, McArthur A, Aromataris E (2018). "Systematic review or scoping review? Guidance for authors when choosing between a systematic or scoping review approach". BMC Medical Research Methodology. 18 (1): 143.
- Nakagawa, Shinichi; Lagisz, Malgorzata; Jennions, Michael D.; Koricheva, Julia; Noble, Daniel W. A.; Parker, Timothy H.; Sánchez-Tójar, Alfredo; Yang, Yefeng; O'Dea, Rose E. (2022). "Methods for testing publication bias in ecological and evolutionary meta-analyses". Methods in Ecology and Evolution. 13 (1): 4–21.
- Newman, Melanie (2019). "Has Cochrane lost its way?". BMJ. 364: k5302.
- Nguyen, Phi-Yen; McKenzie, Joanne E.; Hamilton, Daniel G.; Moher, David; Tugwell, Peter; Fidler, Fiona M.; Haddaway, Neal R.; Higgins, Julian P. T.; Kanukula, Raju; Karunananthan, Sathya; Maxwell, Lara J.; McDonald, Steve; Nakagawa, Shinichi; Nunan, David; Welch, Vivian A. (2023). "Systematic reviewers' perspectives on sharing review data, analytic code, and other materials: A survey". Cochrane Evidence Synthesis and Methods. 1 (2).
- Noma H (2011). "Confidence intervals for a random-effects meta-analysis based on Bartlett-type corrections". Statistics in Medicine. 30 (28): 3304–3312.

- Norris SL, Rehfuess EA, Smith H, Tunçalp Ö, Grimshaw JM, Ford NP, Portela A (1 2019). "Complex health interventions in complex systems: improving the process and methods for evidence-informed health decisions". BMJ Global Health. 4 (Suppl 1): e000963.
- Nunn J, Shafee T, Chang S, Stephens R, Elliott J, Oliver S, John D, Smith M, Orr N. "Standardised Data on Initiatives - STARDIT: Alpha Version". osf.io.
- Nunn JS, Shafee T, Chang S, Stephens R, Elliott J, Oliver S, et al. (2022). "Standardised data on initiatives- STARDIT: Beta version". Research Involvement and Engagement. 8 (1): 31.
- Nunn JS, Tiller J, Fransquet P, Lacaze P (2019). "Public Involvement in Global Genomics Research: A Scoping Review". Frontiers in Public Health. 7: 79.
- Oakley A (2011). A critical woman: Barbara Wootton, social science and public policy in the twentieth century. London: Bloomsbury Academic. ISBN 9781849664707.
- Ortiz, Andrés Felipe Herrera; Camacho, Eduard Cadavid; Rojas, Julián Cubillos; Camacho, Tatiana Cadavid; Guevara, Stephani Zoe; Cuenca, Nury Tatiana Rincón; Perdomo, Andrés Vásquez; Herazo, Valeria Del Castillo; Malo, Rubén Giraldo (2021). "A Practical Guide to Perform a Systematic Literature Review and Meta-analysis". Principles and

- Practice of Clinical Research. 7 (4): 47–57.
- Page MJ, McKenzie JE, Kirkham J, Dwan K, Kramer S, Green S, Forbes A (2014). "Bias due to selective inclusion and reporting of outcomes and analyses in systematic reviews of randomised trials of healthcare interventions". The Cochrane Database of Systematic Reviews. 2015 (10): MR000035.
- Papaioannou D, Sutton A, Carroll C, Booth A, Wong R (2010). "Literature searching for social science systematic reviews: consideration of a range of search techniques". Health Information and Libraries Journal. 27 (2): 114–122.
- Peters MD, Godfrey CM, Khalil H, McInerney P, Parker D, Soares CB (2015). "Guidance for conducting systematic scoping reviews". International Journal of Evidence-Based Healthcare. 13 (3): 141–146.
- Petticrew M, Roberts H (2006). Systematic reviews in the social sciences (PDF). Wiley Blackwell. ISBN 978-1-4051-2110-1. Archived from the original (PDF) on 16 2015.
- Pidgeon TE, Wellstead G, Sagoo H, Jafree DJ, Fowler AJ, Agha RA (2016). "An assessment of the compliance of systematic review articles published in craniofacial surgery with the PRISMA statement guidelines: A systematic review". Journal of Cranio-Maxillo-Facial Surgery. 44 (10): 1522–1530.

- Polanin JR, Tanner-Smith EE, Hennessy EA (2016). "Estimating the Difference Between Published and Unpublished Effect Sizes: A Meta-Review". Review of Educational Research. 86 (1): 207–236.
- Pollock A, Campbell P, Struthers C, Synnot A, Nunn J, Hill S, et al. (21 2017). "Stakeholder involvement in systematic reviews: a protocol for a systematic review of methods, outcomes and effects". Research Involvement and Engagement. 3 (1): 9.
- Pollock A, Campbell P, Struthers C, Synnot A, Nunn J, Hill S, et al. (2018). "Stakeholder involvement in systematic reviews: a scoping review". Systematic Reviews. 7 (1): 208.
- Pollock A, Campbell P, Struthers C, Synnot A, Nunn J, Hill S, et al. (2019). "Development of the ACTIVE framework to describe stakeholder involvement in systematic reviews" (PDF). Journal of Health Services Research & Policy. 24 (4): 245–255.
- Poole C, Greenland S (1999). "Random-effects meta-analyses are not always conservative". American Journal of Epidemiology. 150 (5): 469–475.
- Pursell E, McCrae N (2020). How to Perform a Systematic Literature Review: a guide for healthcare researchers, practitioners and students. Springer Nature. ISBN 978-3-030-49672-2. OCLC 1182880684.

- Quintana, Daniel S. (8 2015). "From pre-registration to publication: a non-technical primer for conducting a meta-analysis to synthesize correlational data". Frontiers in Psychology. 6: 1549.
- Rad, Arian Arjomandi; Nia, Peyman Sardari; Athanasiou, Thanos (2023). "ChatGPT: revolutionizing cardiothoracic surgery research through artificial intelligence". Interdisciplinary CardioVascular and Thoracic Surgery. 36 (6).
- Reddy SM, Patel S, Weyrich M, Fenton J, Viswanathan M (2020). "Comparison of a traditional systematic review approach with review-of-reviews and semi-automation as strategies to update the evidence". Systematic Reviews. 9 (1): 243. doi:10.1186/s13643-020-01450-2. PMC 7574591. PMID 33076975.
- Rethlefsen ML, Farrell AM, Osterhaus Trzasko LC, Brigham TJ (2015). "Librarian co-authors correlated with higher quality reported search strategies in general internal medicine systematic reviews". Journal of Clinical Epidemiology. 68 (6): 617–626.
- Rethlefsen ML, Kirtley S, Waffenschmidt S, Ayala AP, Moher D, Page MJ, Koffel JB (2021). "PRISMA-S: an extension to the PRISMA statement for reporting literature searches in systematic reviews". Journal of the Medical Library Association. 109 (2): 174–200.

- Rethlefsen ML, Murad MH, Livingston EH (2014). "Engaging medical librarians to improve the quality of review articles". JAMA. 312 (10): 999–1000.
- Review Manager (RevMan) [Computer program]. Version 5.2. Copenhagen: The Nordic Cochrane Centre, The Cochrane Collaboration, 2012.
- Riley RD, Ahmed I, Debray TP, Willis BH, Noordzij JP, Higgins JP, Deeks JJ (2015). "Summarising and validating test accuracy results across multiple studies for use in clinical practice". Statistics in Medicine. 34 (13): 2081–2103.
- Riley RD, Higgins JP, Deeks JJ (2011). "Interpretation of random effects meta-analyses". BMJ. 342: d549.
- Roberts I (12 2015). "Retraction of Scientific Papers for Fraud Or Bias Is Just The Tip Of The Iceberg". IFL Science
- Roberts I, Ker K, Edwards P, Beecher D, Manno D, Sydenham E (2015). "The knowledge system underpinning healthcare is not fit for purpose and must change" (PDF). BMJ. 350: h2463.
- Rooney AA, Boyles AL, Wolfe MS, Bucher JR, Thayer KA (2014). "Systematic review and evidence integration for literature-based environmental health science assessments". Environmental Health Perspectives. 122 (7): 711–718.
- Rosenthal R (1979). "The "File Drawer" Problem" and the Tolerance for Null Results". Psychological Bulletin. 86 (3): 638–641.

- Royston P, Parmar MK, Sylvester R (2004). "Construction and validation of a prognostic model across several studies, with an application in superficial bladder cancer". Statistics in Medicine. 23 (6): 907–926.
- Sanderson, S.; Tatt, I. D; Higgins, J. P. (2007). "To> 2013). "Issues in performing a network meta-analysis". Statistical Methods in Medical Research. 22 (2): 169–189.
- Shah HM, Chung KC (2009). "Archie Cochrane and his vision for evidence-based medicine". Plastic and Reconstructive Surgery. 124 (3): 982–988.
- Shapiro S (1994). "Meta-analysis/Shmeta-analysis". American Journal of Epidemiology. 140 (9): 771–778.
- Sharpe, Donald; Poets, Sarena (2020). "Meta-analysis as a response to the replication crisis". Canadian Psychology / Psychologie Canadienne. 61 (4): 377–387.
- Shea BJ, Grimshaw JM, Wells GA, Boers M, Andersson N, Hamel C, et al. (2007). "Development of AMSTAR: a measurement tool to assess the methodological quality of systematic reviews". BMC Medical Research Methodology. 7 (1): 10.
- Shea BJ, Hamel C, Wells GA, Bouter LM, Kristjansson E, Grimshaw J, et al. (2009). "AMSTAR is a reliable and valid

- measurement tool to assess the methodological quality of systematic reviews". Journal of Clinical Epidemiology. 62 (10): 1013-1020.
- Shea BJ, Reeves BC, Wells G, Thuku M, Hamel C, Moran J, et al. (2017). "AMSTAR 2: a critical appraisal tool for systematic reviews that include randomised or non-randomised studies of healthcare interventions, or both". BMJ. 358: j4008.
- Shields, Patricia; Rangarjan, Nandhini (2013). A Playbook for Research Methods: Integrating Conceptual Frameworks and Project Management. Stillwater, Oklahoma: New Forums Press. ISBN 978-1-58107-247-1.
- Shojania KG, Sampson M, Ansari MT, Ji J, Doucette S, Moher D (2007). "How quickly do systematic reviews go out of date? A survival analysis". Annals of Internal Medicine. 147 (4): 224-233.
- Sidik K, Jonkman JN (2002). "A simple confidence interval for meta-analysis". Statistics in Medicine. 21 (21): 3153-3159.
- Siemieniuk R, Guyatt G. "What is GRADE?". BMJ Best Practice (2020).
- Silva V, Grande AJ, Carvalho AP, Martimbianco AL, Riera R (2015). "Overview of systematic reviews - a new type of study. Part II". Sao Paulo Medical Journal = Revista Paulista de Medicina. 133 (3): 206-217.
- Silva V, Grande AJ, Martimbianco AL, Riera R, Carvalho AP (2012). "Overview of systematic reviews - a new type of

study: part I: why and for whom?". Sao Paulo Medical Journal = Revista Paulista de Medicina. 130 (6): 398–404.
- Simmons JP, Nelson LD, Simonsohn U (2011). "False-positive psychology: undisclosed flexibility in data collection and analysis allows presenting anything as significant". Psychological Science. 22 (11): 1359–1366.
- Slavin RE (1986). "Best-Evidence Synthesis: An Alternative to Meta-Analytic and Traditional Reviews". Educational Researcher. 15 (9): 5–9. 0000
- Smith, Mary L.; Glass, Gene V. (1977). "Meta-analysis of psychotherapy outcome studies". American Psychologist. 32 (9): 752–760.
- Song F, Parekh S, Hooper L, Loke YK, Ryder J, Sutton AJ, et al. (2010). "Dissemination and publication of research findings: an updated review of related biases". Health Technology Assessment. 14 (8): iii, ix–xi, 1–193.
- Sood, Amit; Erwin, Patricia J.; Ebbert, Jon O. (2004). "Using Advanced Search Tools on PubMed for Citation Retrieval". o Clinic Proceedings. 79 (10): 1295–1300.
- Synnot A, Bragge P, Lowe D, Nunn JS, O'Sullivan M, Horvat L, et al. (2018). "Research priorities in health communication and participation: international survey of consumers and other stakeholders". BMJ Open. 8 (5): e019481.

- Synnot AJ, Tong A, Bragge P, Lowe D, Nunn JS, O'Sullivan M, et al. (2019). "Selecting, refining and identifying priority Cochrane Reviews in health communication and participation in partnership with consumers and other stakeholders". Health Research Policy and Systems. 17 (1): 45.
- Systematic reviews: CRD's guidance for undertaking reviews in health care (PDF). York: University of York, Centre for Reviews and Dissemination. 2008. ISBN 978-1-900640-47-3.
- The Cochrane Library. 2015 impact factor. Cochrane Database of Systematic Reviews (CDSR) Archived 2 2016 at the Wayback Machine
- Thompson Coon J, Gwernan-Jones R, Garside R, Nunns M, Shaw L, Melendez-Torres GJ, Moore D (2020). "Developing methods for the overarching synthesis of quantitative and qualitative evidence: The interweave synthesis approach". Research Synthesis Methods. 11 (4): 507–521.
- Toews LC (2017). "Compliance of systematic reviews in veterinary journals with Preferred Reporting Items for Systematic Reviews and Meta-Analysis (PRISMA) literature search reporting guidelines". Journal of the Medical Library Association. 105 (3): 233–239.
- Topor, Marta; Pickering, Jade S.; Mendes, Ana Barbosa; Bishop, Dorothy V. M.;

Büttner, Fionn; Elsherif, Mahmoud M.; Evans, Thomas R.; Henderson, Emma L.; Kalandadze, Tamara; Nitschke, Faye T.; Staaks, Janneke P. C.; Akker, Olmo R. van den; Yeung, Siu Kit; Zaneva, Mirela; Lam, Alison (2023). "An integrative framework for planning and conducting Non-Intervention, Reproducible, and Open Systematic Reviews (NIRO-SR)". Meta-Psychology. 7.
- Tranfield D, Denyer D, Smart P (2003). "Towards a methodology for developing evidence-informed management knowledge by means of systematic review". British Journal of Management. 14 (3): 207–222.
- Tricco AC, Lillie E, Zarin W, O'Brien KK, Colquhoun H, Levac D, et al. (2018). "PRISMA Extension for Scoping Reviews (PRISMA-ScR): Checklist and Explanation" (PDF). Annals of Internal Medicine. 169 (7): 467–473.
- Tsafnat G, Glasziou P, Choong MK, Dunn A, Galgani F, Coiera E (2014). "Systematic review automation technologies". Systematic Reviews. 3 (1): 74.
- van Valkenhoef G, Lu G, de Brock B, Hillege H, Ades AE, Welton NJ (2012). "Automating network meta-analysis". Research Synthesis Methods. 3 (4): 285–299.
- Vandvik PO, Brandt L (2020). "Future of Evidence Ecosystem Series: Evidence ecosystems and learning health

- systems: why bother?". Journal of Clinical Epidemiology. 123: 166-170.
- Vevea JL, Woods CM (2005). "Publication bias in research synthesis: sensitivity analysis using a priori weight functions". Psychological Methods. 10 (4): 428-443.
- Viechtbauer, Wolfgang (2010). "Conducting Meta-Analyses in R with the metafor Package". Journal of Statistical Software. 36 (3).
- Vincent, Beatriz; Vincent, Maurice; Ferreira, Carlos Gil (2006). "Making PubMed Searching Simple: Learning to Retrieve Medical Literature Through Interactive Problem Solving". The Oncologist. 11 (3): 243-251.
- Wagner, Gerit; Lukyanenko, Roman; Paré, Guy (2022). "Artificial intelligence and the conduct of literature reviews". Journal of Information Technology. 37 (2): 209-226.
- Whaley P, Aiassa E, Beausoleil C, Beronius A, Bilotta G, Boobis A, et al. (2020). "Recommendations for the conduct of systematic reviews in toxicology and environmental health research (COSTER)". Environment International. 143: 105926.
- White IR (2011). "Multivariate random-effects meta-regression: updates to mvmeta". The Stata Journal. 11 (2): 255-270.
- Whitin (2016). "ROBIS: A new tool to assess risk of bias in systematic reviews

was developed". Journal of Clinical Epidemiology. 69: 225-234.
- ❖ Whiting PF, Rutjes AW, Westwood ME, Mallett S, Deeks JJ, Reitsma JB, et al. (2011). "QUADAS-2: a revised tool for the quality assessment of diagnostic accuracy studies". Annals of Internal Medicine. 155 (8): 529-536.
- ❖ Willis BH, Hyde CJ (2015). "What is the test's accuracy in my practice population? Tailored meta-analysis provides a plausible estimate". Journal of Clinical Epidemiology. 68 (8): 847-854.
- ❖ Willis BH, Hyde CJ (2014). "Estimating a test's accuracy using tailored meta-analysis-How setting-specific data aid study selection". Journal of Clinical Epidemiology. 67 (5): 538-546.
- ❖ Willis BH, Riley RD (2017). "Measuring the statistical validity of summary meta-analysis and meta-regression results for use in clinical practice". Statistics in Medicine. 36 (21): 3283-3301.
- ❖ Wong G, Greenhalgh T, Westhorp G, Buckingham J, Pawson R (2013). "RAMESES publication standards: meta-narrative reviews". Journal of Advanced Nursing. 69 (5): 987-1004.
- ❖ Wong G, Greenhalgh T, Westhorp G, Buckingham J, Pawson R (2013). "RAMESES publication standards: realist syntheses". Journal of Advanced Nursing. 69 (5): 1005-1022.
- ❖ Woodruff TJ, Sutton P (2014). "The Navigation Guide systematic review

methodology: a rigorous and transparent method for translating environmental health science into better health outcomes". Environmental Health Perspectives. 122 (10): 1007–1014.
- ❖ Yates, F.; Cochran, W. G. (1938). "The analysis of groups of experiments". The Journal of Agricultural Science. 28 (4): 556–580.
- ❖ Yoshii A, Plaut DA, McGraw KA, Anderson MJ, Wellik KE (2009). "Analysis of the reporting of search strategies in Cochrane systematic reviews". Journal of the Medical Library Association. 97 (1): 21–29.
- ❖ Zhang H, Deng L, Schiffman M, Qin J, Yu K (2020). "Generalized integration model for improved statistical inference by leveraging external summary data". Biometrika. 107 (3): 689–703.

Annex-1

Sources of published literature

Sources	Literature
Medline	Three different versions: PubMed, OVID Medline and EBSCO Medline (Books at OVID, Full text, Ahead of print, In-Process, I-Data review & other Non-Indexed Citations)
Cochrane Library Cochrane Reviews Cochrane Protocol Other reviews Trials	Intervention and diagnostic reviews Critically appraised and re-structured abstracts Register of clinical trials
EMBASE	Pharmacological literature, conference abstracts Medical devices
Web of Science	Conference abstracts, citation searching, social sciences, Education, Conference Proceedings Citation Index, Science Conference Proceedings Citation Index – Social Science & Humanities
Campbell collaboration	Science/ Social sciences
SCOPUS	Conference abstracts, citation searching, patents, scientific webpages, Trade Publications, Book series
Clinicaltrials.gov Clinical Trials Registry of India (CTRI) South Asian Database of Controlled Clinical Trials (SADCCT)	Trials registered in US and global Trials registered in India Controlled clinical trials and their sources in India and other South Asian countries

Annex-2

Sources of Subject/Study Dependent published literature

Sources	Literature
CINAHL	Nursing and allied health
Psychinfo	Psychology & psychiatry
ERIC	Education
TOXLINE	Effects of drugs and chemicals
PedRO	Physiotherapy (randomized controlled trials and systematic reviews only)
PEDE (Paediatric Economic Database Evaluation)	Paediatric economic evaluations inventory of health state utility weights reported in cost-utility analyses
CEA Registry	Cost-utility analyses on a wide variety of diseases and treatments

Annex 3

Data extraction tool

Data Extraction Form: General Information		
Author(s):		
Systematic Review Title / Reference:		
Publication date:		
Study Background		
Aims / Objectives:		
Variances in Baseline Characteristics to Note:		
Participant Selection Basis:		
Participants		
Index	Baseline	Follow-Up
Controls (count)		
Interventions (count)		
Gender		
Mean Age		
Scope of Age Range		
Ethnicity		
Other Relevant Factors		
Study Details		
Setting / Context:		
Sources Used:		
Appraisal Instruments Used:		
Method(s) of Analysis:		
Outcomes Assessed:		
Outcomes		
Outcome	Units / Count	Definition
Measure of Characteristic A		
Measure of Characteristic B		
Measure of Characteristic C		
Measure of Characteristic D and so on (add rows)		
Data Results		
Outcome:		
Mean Variance:		
Ratio:		
Comments:		

A basic data extraction form (modify as per requirements)

Data extraction tool

1. Data collection form for intervention reviews: RCTs and non-RCTs fn
2. Data extraction form for mapping fn

Data Extraction Form for Randomized Controlled Trials (RCTs)

1. Purpose: To extract detailed information specific to RCTs, focusing on methodology, interventions, and outcomes.
2. Fields:
3. Study ID
4. Author(s)
5. Year of publication
6. Study design: Randomized Controlled Trial (RCT)
7. Randomization method
8. Blinding (participants, personnel, outcome assessors)
9. Allocation concealment
10. Sample size (total and per group)
11. Population characteristics
12. Age
13. Gender
14. Inclusion criteria
15. Exclusion criteria
16. Interventions
17. Description of intervention(s)
18. Duration and frequency of intervention(s)

19. Control/comparator details
20. Outcomes
21. Primary outcomes
22. Secondary outcomes
23. Measurement tools/instruments used
24. Time points of measurement
25. Results
26. Outcome data (e.g., means, standard deviations, confidence intervals)
27. Statistical significance
28. Effect sizes
29. Adverse events
30. Funding source
31. Conflicts of interest
32. Notes/comments

Data Extraction Form for Observational Studies

1. Purpose: To capture relevant data from cohort, case-control, and cross-sectional studies.
2. Fields:
3. Study ID
4. Author(s)
5. Year of publication
6. Study design (e.g., cohort, case-control, cross-sectional)
7. Study period
8. Setting (e.g., hospital, community)
9. Population characteristics
10. Demographics (age, gender, ethnicity)
11. Health status

12. Inclusion criteria
13. Exclusion criteria
14. Exposure(s) or intervention(s)
15. Description of exposure/intervention
16. Duration and frequency
17. Comparator(s)
18. Outcomes
19. Primary outcomes
20. Secondary outcomes
21. Measurement methods
22. Follow-up duration
23. Results
24. Incidence/prevalence rates
25. Risk ratios, odds ratios, hazard ratios
26. Confidence intervals
27. Adjustments for confounding variables
28. Confounding factors
29. Funding source
30. Conflicts of interest
31. Notes/comments

Data Extraction Form for Qualitative Studies

1. Purpose: To extract and summarize qualitative data, focusing on themes, methodologies, and participant experiences.
2. Fields:
3. Study ID
4. Author(s)
5. Year of publication

6. Study design (e.g., phenomenology, grounded theory, ethnography)
7. Research question/aims
8. Setting
9. Population characteristics
10. Demographics (age, gender, ethnicity)
11. Health status
12. Inclusion criteria
13. Exclusion criteria
14. Data collection methods (e.g., interviews, focus groups, observations)
15. Duration of data collection
16. Data analysis methods (e.g., thematic analysis, content analysis)
17. Key findings/themes
18. Participant quotes
19. Researcher's interpretation
20. Study limitations
21. Funding source
22. Conflicts of interest
23. Notes/comments

Data Extraction Form for Systematic Reviews and Meta-Analyses

1. Purpose: To extract data from included systematic reviews and meta-analyses for further synthesis.
2. Fields:
3. Study ID
4. Author(s)
5. Year of publication

6. Type of review (e.g., systematic review, meta-analysis)
7. Research question/aims
8. Number of studies included
9. Study designs of included studies
10. Population characteristics of included studies
11. Interventions/exposures
12. Comparators
13. Outcomes
14. Primary outcomes
15. Secondary outcomes
16. Measurement methods
17. Summary of findings
18. Effect sizes
19. Heterogeneity (I^2 statistic)
20. Confidence intervals
21. P-values
22. Quality assessment of included studies
23. Limitations of the review
24. Funding source
25. Conflicts of interest
26. Notes/comments

Customizing Data Extraction Forms

While the above forms provide a template, it's crucial to customize data extraction forms to suit the specific needs of your systematic review. Consider the following when customizing your form:

Specific Research Question: Tailor the fields to capture data that directly address your research question.

Type of Studies Included: Adjust fields based on whether you are including RCTs, observational studies, qualitative studies, or a mix.

Outcomes of Interest: Ensure the form captures all relevant outcomes and their measurement methods.

Methodological Details: Include fields for critical appraisal of study quality and risk of bias.

Pilot Testing: Test the form with a few studies to identify any necessary adjustments.

Annex 4

PRISMA guidelines

The Preferred Reporting Items for Systematic Reviews and Meta-Analyses (PRISMA) guidelines are a set of evidence-based minimum standards for reporting systematic reviews and meta-analyses. These guidelines help ensure transparency, clarity, and completeness in reporting. PRISMA has evolved over time and has developed extensions to cater to different types of reviews and specific aspects of the review process. Below is an overview of different PRISMA guidelines and extensions:

1. PRISMA 2020

The PRISMA 2020 is the latest update to the original PRISMA statement, which was first published in 2009. It includes a 27-item checklist and an expanded flow diagram.

Key Components:

Title: Indicate that the report is a systematic review, meta-analysis, or both.

Abstract: Structured summary including background, objectives, data sources, study eligibility criteria, methods, results, and conclusions.

Introduction: Rationale and objectives.

Methods: Eligibility criteria, information sources, search strategy, selection process, data collection process, data items, study risk of bias assessment, effect measures, synthesis methods, reporting bias assessment, certainty assessment.

Results: Study selection, study characteristics, risk of bias in studies, results of individual studies, results of syntheses, reporting biases, additional analyses.

Discussion: Summary of evidence, limitations, conclusions.

Funding: Sources of funding and other support.

2. PRISMA-P (Preferred Reporting Items for Systematic Review and Meta-Analysis Protocols)

PRISMA-P is designed for systematic review protocols, guiding researchers on what information to include in their review protocol.

Key Components:

Administrative Information: Title, registration, amendments, support, roles and contributions, amendments.

Introduction: Rationale, objectives.

Methods: Eligibility criteria, information sources, search strategy, study records, data items, risk of bias assessment, data synthesis, meta-bias(es), confidence in cumulative evidence.

3. PRISMA-ScR (Preferred Reporting Items for Systematic Reviews and Meta-Analyses Extension for Scoping Reviews)

PRISMA-ScR provides guidance for scoping reviews, which aim to map the evidence on a particular topic and identify main concepts, theories, sources, and knowledge gaps.

Key Components:

Title: Identify the report as a scoping review.

Abstract: Structured summary including background, objectives, eligibility criteria, information sources, charting methods, results, and conclusions.

Introduction: Rationale, objectives.

Methods: Protocol and registration, eligibility criteria, information sources, search, selection of sources of evidence, data charting process, data items, critical appraisal of individual sources of evidence, synthesis of results.

Results: Selection of sources of evidence, characteristics of sources of evidence, results of individual sources of evidence, synthesis of results.

Discussion: Summary of evidence, limitations, conclusions, and implications.

4. PRISMA-S (Preferred Reporting Items for Systematic Reviews and Meta-Analyses Extension for Searching)

PRISMA-S is focused on improving the reporting of the search component in systematic reviews.

Key Components:

Search Information: Databases and information sources, search strategies, limits and restrictions, updating searches, search management, and peer review of the search strategy.

5. PRISMA-DTA (Preferred Reporting Items for Systematic Reviews and Meta-Analyses of Diagnostic Test Accuracy Studies)

PRISMA-DTA provides guidance for reporting systematic reviews and meta-analyses of diagnostic test accuracy studies.

Key Components:

Title: Identify the report as a systematic review/meta-analysis of diagnostic test accuracy studies.

Abstract: Structured summary including background, objectives, data sources, study eligibility criteria, participants, tests, methods, results, and conclusions.

Introduction: Rationale, objectives.

Methods: Eligibility criteria, information sources, search strategy, selection process, data collection process, data items, risk of bias and applicability, statistical analysis and data synthesis.

Results: Study selection, study characteristics, quality assessment, results of individual studies, synthesis of results, risk of bias across studies, additional analyses.

Discussion: Summary of evidence, limitations, conclusions.

6. PRISMA-NMA (Preferred Reporting Items for Systematic Reviews and Meta-Analyses of Network Meta-Analyses)

PRISMA-NMA provides guidance for reporting systematic reviews that include network meta-analyses, allowing for the comparison of multiple interventions simultaneously.

Key Components:

Title: Identify the report as a systematic review incorporating network meta-analysis.

Abstract: Structured summary including background, objectives, data sources, eligibility criteria, participants, interventions, comparators, outcome measures, results, and conclusions.

Introduction: Rationale, objectives.

Methods: Eligibility criteria, information sources, search strategy, study selection, data collection process, data items, geometry of the network, risk of bias assessment, summary measures, synthesis methods, assessment of inconsistency, exploration of heterogeneity, additional analyses.

Results: Study selection, study characteristics, risk of bias within studies, results of individual studies, synthesis of results, assessment of inconsistency, risk of bias across studies, additional analyses.

Discussion: Summary of evidence, limitations, conclusions.

7. PRISMA-Harms (Preferred Reporting Items for Systematic Reviews and Meta-Analyses of Harms)

PRISMA-Harms extension focuses on reporting harms in systematic reviews of interventions.

Key Components:

Title: Identify the report as a systematic review/meta-analysis focusing on harms.

Abstract: Structured summary including background, objectives, data sources, eligibility criteria, participants, interventions, outcome measures related to harms, results, and conclusions.

Introduction: Rationale, objectives.

Methods: Eligibility criteria, information sources, search strategy, study selection, data collection process, data items, risk of bias in individual studies, synthesis of results.

Results: Study selection, study characteristics, risk of bias within studies, results of individual studies, synthesis of results, risk of bias across studies.

Discussion: Summary of evidence, limitations, conclusions.

8. PRISMA-CI (Preferred Reporting Items for Systematic Reviews and Meta-Analyses of Complex Interventions)

PRISMA-CI focuses on reporting systematic reviews of complex interventions, which involve multiple interacting components.

Key Components:

Title: Identify the report as a systematic review/meta-analysis of complex interventions.

Abstract: Structured summary including background, objectives, data sources, eligibility criteria, participants, interventions, comparators, outcome measures, results, and conclusions.

Introduction: Rationale, objectives.

Methods: Eligibility criteria, information sources, search strategy, study selection, data collection process, data items, complexity dimensions, risk of bias assessment, synthesis methods.

Results: Study selection, study characteristics, risk of bias within studies, results of individual studies, synthesis of results, assessment of heterogeneity, risk of bias across studies.

Discussion: Summary of evidence, limitations, conclusions.

9. PRISMA-Equity: A Reporting Guideline for Systematic Reviews with Equity Considerations

PRISMA-Equity is a specialized extension of the PRISMA guidelines aimed at improving the reporting of systematic reviews that include considerations of health equity. Health equity refers to the absence of avoidable or remediable differences among groups of people, whether those groups are defined socially, economically, demographically, or geographically. PRISMA-Equity provides specific guidance for researchers to systematically and transparently report the equity aspects of their systematic reviews.

The Need for PRISMA-Equity

Systematic reviews often include studies conducted in diverse populations with varying socioeconomic statuses, races, genders, and geographic locations. PRISMA-Equity ensures that such differences are systematically considered and reported, which helps in understanding how interventions impact different population groups and informs equitable health policies and practices.

Key Components of PRISMA-Equity

PRISMA-Equity includes additional items and modifications to the standard PRISMA checklist to address equity-specific issues. Below are the main components and suggested fields that should be included when using PRISMA-Equity:

Title: Indicate that the systematic review addresses health equity considerations.

Abstract: Structured summary including background, objectives, data sources, eligibility criteria, participants, interventions, outcomes, equity considerations, results, and conclusions.

Introduction:

Rationale: Explain the importance of the systematic review in the context of health equity.

Objectives: State the specific objectives related to health equity, such as assessing differential effects of interventions across various population subgroups.

Methods:

Eligibility Criteria: Define criteria for including studies that consider equity factors such as age, gender, ethnicity, socioeconomic status, and geographic location.

Information Sources: Describe sources searched and date of the last search, including databases and grey literature sources that might include equity-focused research.

Search Strategy: Provide a detailed search strategy, including terms used to identify equity-related studies.

Study Selection: Outline the process for selecting studies, emphasizing criteria related to equity.

Data Collection Process: Describe how data were extracted from reports, specifically noting information related to equity factors.

Data Items: List and define all variables for which data were sought, including those related to equity (e.g., subgroup analyses by demographic factors).

Risk of Bias in Individual Studies: Describe methods used to assess risk of bias that may be related to equity factors.

Summary Measures: State the principal summary measures, considering how these may vary across different populations.

Synthesis Methods: Explain methods used to synthesize data, including how equity factors were considered in the synthesis.

Results:

Study Selection: Provide a flow diagram showing the number of studies screened, assessed for eligibility, and included in the review, highlighting equity considerations.

Study Characteristics: Summarize key characteristics of included studies, particularly those related to equity (e.g., population demographics).

Risk of Bias Within Studies: Present data on the risk of bias for each study, including potential biases related to equity.

Results of Individual Studies: Report the findings of individual studies, with a focus on outcomes stratified by equity factors.

Synthesis of Results: Summarize the overall findings, specifically addressing how interventions impacted different population subgroups.

Reporting Biases: Discuss the potential for reporting biases, particularly in relation to equity.

Discussion:

Summary of Evidence: Summarize the main findings with an emphasis on equity-related results.

Limitations: Discuss limitations of the review process and included studies, particularly those related to equity.

Conclusions: State conclusions considering the impact on different population subgroups and implications for health equity.

Sources of Funding: Disclose funding sources and other support, noting any potential conflicts of interest related to equity.

Conflicts of Interest: Disclose any conflicts of interest among the reviewers, particularly those that might influence equity-related assessments.

Implementing PRISMA-Equity: To effectively implement PRISMA-Equity in a systematic review, researchers should:

Integrate Equity from the Start: Consider equity factors in the review question, inclusion criteria, and search strategy.

Use Comprehensive Search Strategies: Include databases and grey literature that may provide information on underrepresented populations.

Systematically Extract Equity Data: Develop a data extraction form that includes fields for equity-related information.

Conduct Subgroup Analyses: Analyze and report findings for different population subgroups to highlight disparities.

Engage with Stakeholders: Involve stakeholders, including representatives from marginalized groups, in the review process to ensure relevant equity issues are addressed.

10. PRISMA for Abstracts 2020

The PRISMA for Abstracts guideline is an extension of the PRISMA 2020 statement, providing specific guidance on how to write informative and comprehensive abstracts for systematic reviews and meta-analyses. Given the importance of abstracts in helping readers quickly ascertain the relevance and findings of a review, PRISMA for Abstracts aims to improve the clarity and completeness of this key section of systematic review reports.

Key Components of PRISMA for Abstracts 2020

PRISMA for Abstracts includes a checklist with essential items that should be included in the abstract of a systematic review or meta-analysis.

Here are the recommended elements, detailed for systematic reviews and meta-analyses:

Title: Identify the report as a systematic review, meta-analysis, or both.

Background: Briefly describe the context or rationale for the review.

Objectives: State the main objectives of the review.

Eligibility Criteria: Specify the criteria for study inclusion (e.g., types of studies, participants, interventions, comparisons, outcomes).

Information Sources: Indicate the main databases and other information sources searched.

Methods: Outline the basic methodology, including the data selection and extraction process.

Risk of Bias: Summarize how risk of bias was assessed in the included studies.

Results: Provide the number of studies included and the main findings, including relevant effect estimates and their precision (e.g., confidence intervals).

Discussion/Conclusions:

Summarize the conclusions, highlighting the main findings and their implications.

11. PRISMA-IPD: Preferred Reporting Items for Systematic Reviews and Meta-Analyses of Individual Participant Data

PRISMA-IPD (Preferred Reporting Items for Systematic Reviews and Meta-Analyses of Individual Participant Data) is a specialized extension of the PRISMA guidelines. It focuses on systematic reviews and meta-analyses that use individual participant data (IPD), which is the raw data collected from each participant in the included studies. PRISMA-IPD aims to enhance transparency, completeness, and consistency in the reporting of IPD systematic reviews and meta-analyses.

Key Components of PRISMA-IPD

The PRISMA-IPD guidelines include modifications and extensions of the original PRISMA checklist items, tailored to the specifics of IPD systematic reviews. Below are the key components and recommended items for PRISMA-IPD:

Title: Identify the report as a systematic review, meta-analysis, or both, and indicate that it involves individual participant data.

Abstract: Provide a structured summary including background, objectives, data sources, eligibility criteria, participants, interventions, comparators, outcomes, study appraisal and synthesis methods, results, limitations, conclusions, and registration number.

Rationale: Explain the reasons for using IPD in the systematic review and the advantages it provides over aggregate data.

Objectives: Clearly state the main objectives of the review, specifying the research questions addressed with IPD.

Eligibility Criteria: Specify criteria for including studies, types of participants, interventions, comparators, and outcomes. Include details about IPD availability and accessibility.

Information Sources: List all data sources, databases, and registries searched to identify studies and obtain IPD. Mention any efforts to contact study authors for additional data.

Search Strategy: Provide the full search strategy for at least one major database, including keywords and search terms used to identify relevant studies.

Data Collection Process: Describe methods used to collect IPD, including procedures for data sharing, acquisition, and validation.

Data Items: Define all variables for which IPD were sought, including demographic, baseline, and outcome variables.

Risk of Bias Assessment: Explain methods used to assess the risk of bias in individual studies and in the IPD itself.

Synthesis Methods: Detail the statistical methods used to synthesize IPD, including approaches for dealing with missing data, subgroup analyses, and meta-regression.

Results: Summarize the characteristics of included studies and participants, main findings, effect estimates, and measures of variability. Report results of risk of bias assessments.

Discussion: Discuss the main findings, their implications for practice and policy, limitations of the review, and suggestions for future research.

Funding: Disclose sources of funding and other support, and note any potential conflicts of interest.

Registration: Provide the registration number and name of the registry where the systematic review protocol is registered.

Annex 5

Critical appraisal tools

Critical appraisal tools are essential for evaluating the quality and reliability of studies included in systematic reviews. These tools help reviewers assess the risk of bias, methodological rigor, and overall validity of the studies, ensuring that the conclusions drawn from the review are based on high-quality evidence. Different tools are tailored for different types of studies, including randomized controlled trials (RCTs), observational studies, diagnostic accuracy studies, and qualitative research. Below are checklists for the different critical appraisal tools, generally used:

1. Cochrane Risk of Bias Tool (RoB 2)

Domain 1: Bias Arising from the Randomization Process

1. Was the allocation sequence random?
2. Was the allocation sequence concealed until participants were assigned to interventions?
3. Were there baseline imbalances that suggest a problem with the randomization process?

Domain 2: Bias Due to Deviations from Intended Interventions

1. Were participants aware of their assigned intervention during the trial?
2. Were those administering the interventions aware of each participant's assigned intervention during the trial?
3. Were deviations from intended intervention balanced between groups?
4. Was an appropriate analysis used to estimate the effect of assignment to intervention?

Domain 3: Bias Due to Missing Outcome Data

1. Were data for all participants included in the analysis?
2. Were missing outcome data balanced between groups?
3. Were missing data related to their true value?

Domain 4: Bias in Measurement of the Outcome

1. Was the method of measuring the outcome appropriate?
2. Were outcome assessors aware of the intervention received by participants?
3. Could measurement of the outcome have been influenced by knowledge of the intervention?

Domain 5: Bias in Selection of the Reported Result

1. Was the reported outcome chosen from multiple measurements?
2. Was the analysis plan pre-specified?

2. Newcastle-Ottawa Scale (NOS)

Selection

1. Is the exposed cohort truly or somewhat representative of the average patient in the community?
2. Was the non-exposed cohort drawn from the same community as the exposed cohort?
3. Was ascertainment of exposure to risks or treatments secure and validated?
4. Demonstration that the outcome of interest was not present at start of study.

Comparability

1. Are the cohorts comparable on the basis of the design or analysis? Controlled for confounders.

Outcome

2. Was the outcome assessed independently or was the assessor blind to the exposure status?
3. Was the follow-up long enough for outcomes to occur?

4. Was follow-up complete, or were patients lost to follow-up adequately described and accounted for?

3. QUADAS-2 (Quality Assessment of Diagnostic Accuracy Studies)

Domain 1: Patient Selection

1. Were the criteria for inclusion clearly described?
2. Was a consecutive or random sample of patients enrolled?
3. Was a case-control design avoided?
4. Did the study avoid inappropriate exclusions?

Domain 2: Index Test

1. Were the index test results interpreted without knowledge of the reference standard?
2. Was the execution of the index test described in sufficient detail to allow replication?

Domain 3: Reference Standard

1. Is the reference standard likely to correctly classify the target condition?
2. Were the reference standard results interpreted without knowledge of the index

Domain 4: Flow and Timing

1. Was there an appropriate interval between the index test and the reference standard?
2. Did all patients receive the same reference standard?
3. Were all patients included in the analysis?

4. CASP (Critical Appraisal Skills Programme)

Section 1: Validity

1. Did the study address a clearly focused question?
2. Was the cohort recruited in an acceptable way?
3. Was the exposure accurately measured to minimize bias?
4. Was the outcome accurately measured to minimize bias?

Section 2: Results

1. What are the results?
2. How precise are the results?
3. Do you believe the results?

Section 3: Applicability

1. Can the results be applied to the local population?

2. Do the results fit with other available evidence?
3. What are the implications of this study for practice?

5. AMSTAR 2 (A Measurement Tool to Assess Systematic Reviews)

Item 1: Protocol

1. Did the research questions and inclusion criteria for the review include the components of PICO?
2. Did the review protocol pre-specify the objectives and methods?

Item 2: Study Selection and Data Extraction

1. Did the review authors explain their selection of the study designs for inclusion?
2. Was study selection and data extraction performed in duplicate?

Item 3: Literature Search

1. Was an adequate literature search performed?

Item 4: Risk of Bias

1. Was the risk of bias assessed for individual studies?
2. Was the impact of risk of bias on results discussed?

Item 5: Data Synthesis

1. Were appropriate methods used to combine the results of studies?
2. Was the potential for publication bias assessed?

Item 6: Funding

1. Was the funding source for the review stated?
2. Were any conflicts of interest declared?

6. ROBINS-I (Risk of Bias in Non-randomized Studies of Interventions)

Domain 1: Bias Due to Confounding

1. Were confounding variables identified?
2. Were strategies to deal with confounding variables stated?

Domain 2: Bias in Selection of Participants

1. Were participants selected into the study in a way that could introduce bias?

Domain 3: Bias in Classification of Interventions

1. Was the intervention status classified correctly?

Domain 4: Bias Due to Deviations from Intended Interventions

1. Were there deviations from the intended interventions?
2. Were these deviations adequately reported?

Domain 5: Bias Due to Missing Data

1. Was the outcome data incomplete?
2. Were the reasons for missing data reported?

Domain 6: Bias in Measurement of Outcomes

1. Was the method of outcome assessment appropriate?
2. Were the assessors of outcome blinded to intervention status?

Domain 7: Bias in Selection of the Reported Result

1. Was the reported result selected from multiple analyses?

7. GRADE (Grading of Recommendations Assessment, Development, and Evaluation)

Quality of Evidence

1. Study limitations: Was there risk of bias?
2. Inconsistency: Were the results consistent across studies?

3. Indirectness: Was there indirect evidence?
4. Imprecision: Were the results precise?
5. Publication bias: Was there evidence of publication bias?

Strength of Recommendations

1. Balance between desirable and undesirable effects.
2. Quality of evidence.
3. Values and preferences.
4. Resource use and cost-effectiveness.

8. MMAT (Mixed Methods Appraisal Tool)

For Qualitative Research:

1. Are the qualitative research questions clearly stated?
2. Is the qualitative approach appropriate to answer the research question?
3. Are the qualitative data collection methods adequate to address the research question?
4. Are the findings supported by the data?

For Quantitative Randomized Controlled Trials:

1. Is the randomization appropriately performed?
2. Are the groups comparable at baseline?
3. Are there complete outcome data?

4. Are all expected outcomes reported?

For Quantitative Non-Randomized Studies:

1. Are the participants representative of the target population?
2. Are measurements appropriate regarding both the outcome and exposure?
3. Are there complete outcome data?
4. Do the statistical analyses control for confounding variables?

For Quantitative Descriptive Studies:

1. Is the sampling strategy relevant to address the research question?
2. Is the sample representative of the target population?
3. Are the measurements appropriate?
4. Is the risk of nonresponse bias low?

For Mixed Methods Studies:

1. Is there an adequate rationale for using a mixed methods design?
2. Are the different components of the study effectively integrated to answer the research question?
3. Are the results adequately brought together to provide a coherent conclusion?
4. Are the outputs of the different components adequately described?

Annex 6

Methods of search in different databases

Suppose we have the topic "Mother-to-child HIV transmission in India"

PubMed Search Strategy

Keywords and MeSH Terms:

1. Mother-to-child HIV transmission
2. Vertical HIV transmission
3. India
4. HIV
5. AIDS
6. Prevention of Mother-to-Child Transmission (PMTCT)

Search Query:

("HIV"[MeSH Terms] OR "HIV Infections"[MeSH Terms] OR HIV[Title/Abstract] OR AIDS[Title/Abstract] OR "Human Immunodeficiency Virus"[Title/Abstract])

AND

("Vertical Transmission"[MeSH Terms] OR "Vertical transmission of infectious disease"[MeSH Terms] OR "Mother-to-Child Transmission"[Title/Abstract] OR "Perinatal

Transmission"[Title/Abstract] OR "Vertical HIV Transmission"[Title/Abstract])

AND

("India"[MeSH Terms] OR India[Title/Abstract] OR Indian[Title/Abstract])

**Adjustment of year through applying of filters

Google Scholar Search Strategy

When searching on Google Scholar for the topic "Mother-to-child HIV transmission in India," you can use a combination of specific keywords and phrases to maximize the relevance of your search results. Here's how you can structure your search query:

Basic Search Query:

"Mother-to-child HIV transmission" AND India

This basic query targets the key topic and geographic focus directly.

Expanded Search Query:

To capture a broader range of relevant articles, include synonyms and related terms:

("Mother-to-child HIV transmission" OR "Vertical HIV transmission" OR "Perinatal HIV transmission" OR "MTCT of HIV") AND (India OR Indian)

*Adjustment of year through applying of filters

EMBASE Search Strategy

Keywords and Emtree Terms:

1. HIV
2. AIDS
3. Vertical transmission
4. Mother-to-child transmission
5. Perinatal transmission
6. India

Structured Search Query:

('human immunodeficiency virus'/exp OR 'HIV infection'/exp OR HIV:ti,ab,kw OR AIDS:ti,ab,kw OR 'human immunodeficiency virus':ti,ab,kw)

AND

('vertical transmission'/exp OR 'mother to child transmission of infectious disease'/exp OR 'mother-to-child transmission':ti,ab,kw OR 'perinatal transmission':ti,ab,kw)

AND

('India'/exp OR India:ti,ab,kw OR Indian:ti,ab,kw)

*Adjustment of year

In EMBASE,

ti: Title

This tag searches for terms within the titles of articles.

ab: Abstract

This tag searches for terms within the abstracts of articles.

kw: Keywords

This tag searches for terms within the keywords provided by the authors.

PsycINFO Search Strategy

Keywords and Subject Terms:

1. Mother-to-child HIV transmission
2. Vertical HIV transmission
3. Perinatal HIV transmission
4. India
5. HIV
6. AIDS

Structured Search Query:

("Mother-to-child HIV transmission" OR "Vertical HIV transmission" OR "Perinatal HIV transmission" OR "MTCT of HIV") AND

(India OR Indian) AND (HIV OR AIDS OR "Human Immunodeficiency Virus")

*Adjustment of year

MEDLINE Search Strategy

Keywords and MeSH Terms:

1. Mother-to-child HIV transmission
2. Vertical HIV transmission
3. Perinatal HIV transmission
4. India
5. HIV
6. AIDS
7. Prevention of Mother-to-Child Transmission (PMTCT)

Structured Search Query:

("HIV"[MeSH Terms] OR "HIV Infections"[MeSH Terms] OR HIV[Title/Abstract] OR AIDS[Title/Abstract] OR "Human Immunodeficiency Virus"[Title/Abstract])

AND

("Vertical Transmission"[MeSH Terms] OR "Vertical Transmission of Infectious Disease"[MeSH Terms] OR "Mother-to-Child Transmission"[Title/Abstract] OR "Perinatal

Transmission"[Title/Abstract] OR "Vertical HIV Transmission"[Title/Abstract])

AND

("India"[MeSH Terms] OR India[Title/Abstract] OR Indian[Title/Abstract])

*Adjustment of year

Web of Science Search Strategy

Keywords:

1. Mother-to-child HIV transmission
2. Vertical HIV transmission
3. Perinatal HIV transmission
4. India
5. HIV
6. AIDS
7. MTCT (Mother-to-Child Transmission)
8. PMTCT (Prevention of Mother-to-Child Transmission)

Structured Search Query:

TS=("Mother-to-child HIV transmission" OR "Vertical HIV transmission" OR "Perinatal HIV transmission" OR "MTCT of HIV") AND

TS=(India OR Indian) AND TS=(HIV OR AIDS OR "Human Immunodeficiency Virus")

***Adjustment of year**

In Web of Science, **TS** stands for Topic Search. When you use TS, you are instructing the database to search for the specified terms within the title, abstract, author keywords, and Keywords Plus of the articles.

Using TS is beneficial because it broadens the search to include all the main text fields where the topic might be discussed, ensuring a comprehensive search. Here is how you can construct a search query using TS for the topic "Mother-to-child HIV transmission in India".

Published book by the author

Cyber-Physical System to Artificial Intelligence in the Era of New Healthcare Supporting Systems, Atlanta Publication, printed in USA, 1st edition (30 March 2023), paperback: p 298. ISBN-13:9798389286580, Hardcover: p 298; ISBN-13: 979-8389288539,

An Essential in Public Health and Epidemiology, Notion Press; 1st edition (27 June 2022), paperback:p 196. ISBN-10: 8887335940, ISBN-13: 979-8887335940,

The Black Fungus: An overview on Mucormycosis, Notion Press; 1st edition (21 October 2021), paperback: p 92. ISBN-13: 978-1684872213.

Hepatitis: Viral, Toxic, Alcoholic & Autoimmune, Notion Press; 1st edition (31 March 2021), paperback: p 286. ISBN-13: 978-1638737872.

Coronaviruses: At a glance, Notion Press; 1st edition (27 January 2021), paperback: p 128. ISBN-13: 978-1638062264.

An Overview of Systematic Review and Meta-analysis, Amazon Press, Atlanta, printed in USA, 1st edition (22 January 2021), paperback: p 168. ISBN-13: 979- 859815470

Mihir Bhatta

Mihir Bhatta (M.Sc., Ph.D., PGDPHM, MPH, FCRSD, FSASS.) is a scientific worker in the field of viral research, public health, and epidemiology, especially in human immunodeficiency virus (HIV), sexually transmitted infections (STI) and viral hepatitis research. Dr. Mihir Bhatta, along with his co-workers published numerous research articles in national and international journals. His research work is based on viral research, computation, Bio-Statistics, GIS techniques, animal science, public health, epidemiology, ecology, and Environment.

He is trained on machine learning and artificial intelligence (AI) in systematic reviews; PRISMA Reporting and application of COVIDENCE to perform evidence synthesis also participate in Cochrane Evidence Synthesis as Cochrane Learning Initiative. Trained in performing meta-analyses in the case of very few studies. Working on developing Search Strategies for Clinical

AUTHOR

Trials under PICOS Framework with Bodleian Health Care Libraries, the University of Oxford, United Kingdom. He also achieved Six Sigma Yellow and White belt. Till date he published several Systematic reviews, many Systematic review and Meta-analysis and very few Protocol of Systematic review. He was participated in the Cochrane's Annual General Meeting 2022 (AGM-2022) at Madrid, Barcelona, Spain. He also written the book "An Overview of Systematic Review and Meta-analysis" along with other books.

About the Book

Synthesizing Evidence: The Art of Systematic Review: A Comprehensive Guide to Evidence Synthesis, Methodology, and Impact

Purpose and Scope

The primary goal of Synthesizing Evidence: The Art of Systematic Review is to offer readers a robust understanding of the systematic review process, from inception to completion. It provides an in-depth exploration of the methodologies involved in evidence synthesis, ensuring that readers are equipped with the skills needed to conduct high-quality reviews. The book also emphasizes the importance of systematic reviews in informing clinical practice, policy-making, and future research directions.

Key Components

Introduction to Systematic Reviews:

The book begins with a foundational overview of systematic reviews, distinguishing them from other types of reviews such as narrative reviews and scoping reviews. It highlights the importance of systematic reviews in evidence-based practice.

Methodological Rigor:

Detailed chapters are dedicated to the methodologies involved in conducting systematic reviews. This includes formulating research questions, developing search strategies, and selecting appropriate databases. The book stresses the importance of using rigorous and reproducible methods to ensure the reliability and validity of the review findings.

Data Extraction and Management:

A comprehensive guide to data extraction is provided, outlining best practices for managing and organizing data. The book covers the use of standardized data extraction forms and software tools to streamline the process.

Critical Appraisal:

The text includes extensive coverage on the critical appraisal of studies, offering tools and checklists to assess the quality of included studies. This section ensures that readers can differentiate between high-quality and low-quality evidence.

Data Synthesis and Analysis:

Readers are introduced to various methods of data synthesis, including meta-analysis for quantitative data and thematic synthesis for qualitative data. The book provides step-by-step instructions and practical examples to facilitate understanding.

Reporting and Dissemination:

Guidance on how to effectively report the findings of a systematic review is provided, emphasizing adherence to reporting guidelines such as PRISMA (Preferred Reporting Items for Systematic Reviews and Meta-Analyses). The book also discusses strategies for disseminating review findings to ensure maximum impact.

Practical Applications

Synthesizing Evidence: The Art of Systematic Review is not only theoretical but also highly practical. It includes case studies, real-world examples, and practical exercises to help readers apply what they have learned. The book also addresses common challenges and pitfalls in conducting systematic reviews, offering solutions and tips for overcoming these obstacles.

Impact on Evidence-Based Practice

The book underscores the critical role of systematic reviews in evidence-based practice. By synthesizing existing research, systematic reviews provide a high level of evidence that can inform clinical guidelines, policy decisions, and future research. The text highlights the impact of systematic reviews on improving healthcare outcomes and advancing scientific knowledge.

Audience

This book is an essential resource for:

Researchers: Both novice and experienced researchers will find valuable insights and detailed guidance on conducting systematic reviews.

Healthcare Professionals: Clinicians and healthcare practitioners can use this book to better understand the evidence that underpins clinical guidelines and practices.

Students: Those studying health sciences, public health, or related fields will benefit from the clear explanations and practical guidance.

Policy Makers: The book provides a robust framework for using systematic reviews to inform policy and decision-making processes.

www.ingramcontent.com/pod-product-compliance
Lightning Source LLC
Chambersburg PA
CBHW050047230526
45470CB00004B/1432